my revisi⏻n notes

OCR GCSE (9–1)
DESIGN & TECHNOLOGY

Andy Knight
Kevin Crampton
Corinne Walkley

HODDER
EDUCATION
AN HACHETTE UK COMPANY

Picture Credits

Every effort has been made to trace and acknowledge ownership of copyright. The publishers will be glad to make suitable arrangements with any copyright holders whom it has not been possible to contact. The authors and publishers would like to thank the following for permission to reproduce copyright material: fig. 3.1 © Dorling Kindersley ltd/Alamy Stock Photo; fig. 7.1 © Martien van Gaalen/123RF; fig. 8.8 © Viktor Bonda/123RF; fig. 10.2 ©Tom Mc Nemar/istockphoto.com; fig. 10.4 © batchimages/Alamy Stock Photo; fig. 10.6 © Steven Heap/123RF; fig. 10.8 © Russell Roberds/istockphoto; fig. 10.9 © AKS – Fotolia; fig. 10.10 © Udom Jinama/123RF; fig. 10.11 © vitaliy_73/Shutterstock.com; fig. 10.12 © artemisphoto/Shutterstock.com; fig. 11.3 © Joe Gough – Fotolia; fig. 11.4 © Matt Blythe/Shutterstock.com; fig. 11.5 © Gail Johnson – Fotolia; fig. 12.1 © Trevor Walker/Alamy Stock Photo; fig. 12.2 © Westend61/Getty Images; fig. 12.3 © Anton Samsonov/123RF; fig. 12.4 © Olegsam/123RF; fig. 13.3 © BMPix/istockphoto.com; fig. 13.4 © cunaplus/stock.adobe.com; fig. 13.7 © Roman Samokhin – Fotolia; fig. 13.8 © Schneider Sebastian/Alamy Stock Photo; fig. 15.1 © Anterovium – Fotolia; fig. 15.5 © Paul Smith/123RF; fig. 15.10 © B Christopher/Alamy Stock Photo; fig 17.5 © rostovdriver/stock.adobe.com; fig. 18.4 © Digital Genetics/Shutterstock.com, © Unkas Photo/Shutterstock.com, © Coprid/Shutterstock.com, © KRIANGKRAI APKARAT/123RF, © Achim Prill/123 RF, © Elena Abduramanova/Shutterstock.com; fig. 18.5 © pzAxe/Shutterstock.com; fig. 18.6 © Happy Stock Photo/Shutterstock.com; fig. 18.7 © Artem Loskutnikov/Shutterstock.com; fig. 18.19 © mariakraynova/Fotolia; fig. 19.1 © Getty Images/iStockphoto/Thinkstock, © Stockbyte/Getty Images/Flowers SD119, © Christopher May/Shutterstock.com, © Tinnko/Shutterstock.com; fig. 19.2 © JoobheadiStock/Getty Images Plus; fig. 19.3 © Coprid/Shutterstock.com; fig. 19.6 © Design Pics Inc/Alamy Stock Photo; fig. 19.7 © Mamunur Rashid/Alamy Stock Photo; fig. 19.9 © Shutter Top/Shutterstock.com; fig. 19.11 © SCIENCE PHOTO LIBRARY; fig. 19.12 © Indigolotos/123 RF; fig. 19.13 © Dorling Kindersley/Shutterstock.com; fig. 19.20 © Dario Cantatore/Stringer/Getty Images Entertainment/Getty Images; fig. 19.21 By Creative Lab; fig. 19.22 © Natalie Behring/Aurora Photos/Alamy Stock Photo; p.160 (top) ©Tom Mc Nemar/istockphoto.com; p.164 © praisaeng/Fotolia; p.167 © Petr Malyshev/Fotolia. The authors and publishers would also like to thank the following schools and students: Silcoates School in Wakefield; Kenilworth School in Kenilworth; Philip Robinson from Thrybergh Academy and Sports College and Terry Bream. Any photos not listed above are copyright of the authors.

Hachette UK's policy is to use papers that are natural, renewable and recyclable products and made from wood grown in well-managed forests and other controlled sources. The logging and manufacturing processes are expected to conform to the environmental regulations of the country of origin.

Orders: please contact Hachette UK Distribution, Hely Hutchinson Centre, Milton Road, Didcot, Oxfordshire, OX11 7HH. Telephone: +44 (0)1235 827827. Email education@hachette.co.uk Lines are open from 9 a.m. to 5 p.m., Monday to Friday. You can also order through our website: www.hoddereducation.co.uk

ISBN: 9781510432284

© Andy Knight, Kevin Crampton, Corinne Walkley 2018

First published in 2018 by
Hodder Education,
An Hachette UK Company
Carmelite House
50 Victoria Embankment
London EC4Y 0DZ
www.hoddereducation.co.uk

Impression number   10 9 8 7 6 5 4
Year   2022   2021

Cover photo © Brain light/Alamy Stock Photo

Typeset by Integra Software Services Pvt. Ltd., Pondicherry, India

Printed by CPI Group (UK) Ltd, Croydon CR0 4YY

A catalogue record for this title is available from the British Library.

Get the most from this book

Everyone has to decide their own revision strategy, but it is essential to review your work, learn key facts and test your understanding. These Revision Notes will help you to do that in a planned way, topic by topic. You can check your progress by ticking off each section as you revise.

Tick to track your progress

Use the revision planner on pages vi–ix to plan your revision, topic by topic. Tick each box when you have:

- revised and understood a topic
- tested yourself
- practised exam questions and gone online to check your answers and complete the quick quizzes.

You can also keep track of your revision by ticking off each topic heading in the book. You may find it helpful to add your own notes as you work through each topic.

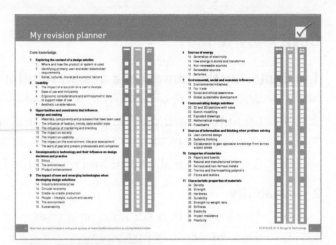

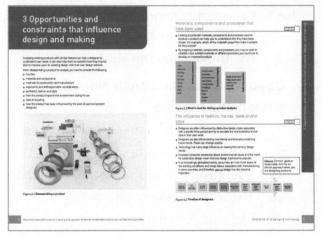

Features to help you succeed

Expert tips are given throughout the book to help you polish your exam technique in order to maximise your chances in the exam.

The authors identify the typical mistakes candidates make and explain how you can avoid them.

These short, knowledge-based questions provide the first step in testing your learning. Answers are given online.

Key terms from the specification are highlighted in bold throughout the book.

Go online to try out the extra quick quizzes at **www.hoddereducation.co.uk/myrevisionnotes**

My revision planner

Core knowledge

REVISED   TESTED   EXAM READY

Now test yourself answers and quick quizzes at **www.hoddereducation.co.uk/myrevisionnotes**

	REVISED	TESTED	EXAM READY

In-depth knowledge

In the exam you will only be expected to demonstrate in-depth knowledge of one of these material categories or systems.

Now test yourself answers and exam practice answers at www.hoddereducation.co.uk/myrevisionnotes

Countdown to my exams

1 Exploring the context of a design solution

A **design solution** is where a product or system fulfils a need or want. A design opportunity happens where there is a gap in the existing market.

- Every living thing has needs and wants.
- These needs and wants are different for everyone.
- What one person needs or wants may be the complete opposite to the needs and wants of the person next to them.
- One person's needs and wants may cause difficulties or problems for another person.
- Each person or living thing has unique and individual needs and wants, but there are common needs that are shared between groups.
- Designers design products or systems to fulfil needs and wants.

> **Design solution:** Where a product or system fulfils a need or want.

Figure 1.1 **What leads to a design solution**

Where and how the product or system is used REVISED

Why design a new product or system?

- All new designs of products and systems should ideally make things easier, quicker or cheaper for a user.

What needs to be considered before designing?

- Designers must consider what issues may arise before designing a product or system.
- A successful design solution will identify, address and take into consideration many of the issues highlighted.
- Potential issues should be identified within the design context.
- Designers need to consider the impact on the environment and other users.

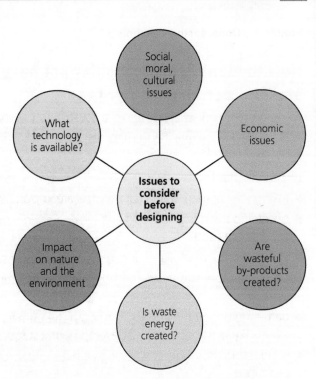

Figure 1.2 **Issues to be considered before designing**

Identifying primary user and wider stakeholder requirements

Most products and systems are used or maintained by humans.

- The **primary user** is the main user of the product or system.
- A **stakeholder** is a person, group or organisation that has an interest in the product or system.

When designing a product or system, the designer must not only consider the views of the primary user and think about how they will interact with the product, but also consider the stakeholders and wider stakeholders that have an interest in the designs they are developing.

> **Primary user:** The main user of the product or system.
>
> **Stakeholder:** A user, person, group or organisation that has an interest in the product or system.

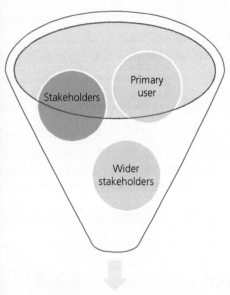

Viewpoints to be considered

Figure 1.3 Considering viewpoints

How to identify and consider primary user and stakeholder requirements

You can use a **task analysis** and the **5WHs** – who, what, when, where, why – to identify who your primary users and wider stakeholders are and their requirements.

You can do this by:

- asking your primary user what they feel are important things to consider
- observing the design context and looking at how similar products/systems are used
- identifying successful and unsuccessful features of similar products
- talking to other stakeholders who may have an interest in the product/system
- gathering ongoing feedback from both the primary user and the stakeholders throughout the developmental stage
- conducting thorough user testing with designs, prototypes and solutions produced
- continuing the iterative process by re-evaluating the product/system design to ensure an optimum solution is reached.

> **Task analysis:** An exploration of the design context.
>
> **5WHs:** A way of identifying the needs and wants of the primary user and stakeholders by asking who, what, when, where and why.

Social, cultural, moral and economic factors

1 Exploring the context of a design solution

To make a product or system successful designers must consider **social**, **cultural**, **moral** and **economic factors**. These factors usually overlap and affect all aspects of our lives and how we interact as humans.

Social factors

These are lifestyle factors that affect people within our society. They involve considerations or problems that can affect people's lives, such as:

- anti social behaviour
- poverty
- drug abuse
- alcohol abuse
- economic deprivation
- unemployment.

Designers have a responsibility to consider these factors when designing.

Designers should aim to design products that can be used by as many people as possible, including people with disabilities and those of different ages (including the elderly). This is called **inclusive design**.

Cultural factors

These are the beliefs, moral values, traditions, language, laws and behaviours that are common to a group of people (for example, a nation or community).

- A product that is suitable for or appealing to one society or culture will not necessarily be successful or useful in another society or culture, and designers need to consider this when designing.
- Designers have a responsibility to recognise, understand and respect different cultural beliefs when designing products. (Colours, for example, have different meanings in different cultures – designers need to be aware of these to ensure their product is successful in a particular society and to avoid giving offence.)

Moral factors

A moral issue is related to human behaviour. It is the distinction between good and bad, or right and wrong, behaviour according to our conscience.

- A designer may not be happy that their design could be used to promote products that are harmful to the consumer. An example of this could be a designer making a label for a sugary drink. Excessive consumption of sugary drinks could result in long-term health problems such as diabetes and obesity.
- Moral factors may influence a designer's choice of materials. For example, using recyclable materials or avoiding unethical materials such as fur.
- The way that we design and make things affects the safety, comfort and well-being of the people who come into contact with them. The morality of our thinking and decision making has an impact on every aspect of our design and technology work, and on the people who will use our products.

Social factors: Lifestyle factors that affect people within our society.

Cultural factors: The beliefs, moral values, traditions, language, laws and behaviours that are common to a group of people (for example, a nation or community).

Moral factors: A moral issue is related to human behaviour; it is the distinction between good and bad, or right and wrong, behaviour, according to our conscience.

Economic factors: How the making, using and disposing of products and services can have an impact on the industry and trade of a country.

Inclusive design: The design of mainstream products and/or services that are accessible to, and usable by, as many people as reasonably possible without the need for special adaptation or specialised design.

OCR GCSE (9–1) Design & Technology   3

Economic factors

Economic factors are concerned with how money is made and used, and whether we are able to supply and maintain enough resources to manufacture the products we design. A country's economy is connected to how many goods or services that country produces and how much money people can spend on them.

- Making, using and disposing of products will have an impact on the economy; it can create or affect jobs in a positive or negative way.
- Will the product or service create or threaten jobs? For example, many modern products are manufactured by computer-controlled systems (CAM), resulting in the loss of jobs for skilled workers in factories.
- While some jobs may be lost (traditional skills, manual labour), some new jobs can be created (developing and maintaining new and emerging technology).
- Many products that we buy are manufactured by people who are badly paid and who work in poor conditions, sometimes abroad. This can have a negative effect on that country's economy as the workers have little money to spend.
- Big brand names in fashion, fast food, drinks, supermarkets, DIY stores, cars and motorbikes can be seen in many parts of the world; this is called product **globalisation**.

> **Globalisation:** The process by which the world is becoming increasingly interconnected as a result of massively increased trade and cultural exchange.

Exam tip

Whenever you have to consider social, cultural, moral and economic factors when answering an exam question, you need to think outside the box. Try to think about the advantages and disadvantages of the issue in relation to these factors; try to look at it from lots of different perspectives relating to age, gender and race, not just your own. All four factors overlap, so if an issue has a moral aspect, it will have social, cultural and economic concerns too.

Typical mistake

When answering an extended response question, you will need to think carefully about how you structure your response to ensure you can access maximum marks. Your response needs to be detailed and show that you are able to understand the wider issues. Think about the advantages and disadvantages of the topic being examined in the question; think about the primary user and how they are affected; consider and discuss how the issues may affect the wider stakeholders. Try to use the correct grammar, technical language and specialist terms too.

Now test yourself

TESTED

1 List three needs or wants for a three-year-old child. [3 marks]
2 Explain the difference between a primary user and a stakeholder. [2 marks]
3 Give one social factor that might affect the primary user's ability to purchase a new product. [1 mark]
4 Explain how moral factors might influence a designer. [3 marks]
5 What term do we use to show that a product is accessible to all users? [1 mark]

2 Usability

The impact of a solution on a user's lifestyle

REVISED

Products can change the way we live and work. **Usability** – how easy a product is to use – can have an impact on our lives. Products that make jobs we dislike doing quicker or easier may be welcomed, for example a dishwasher can make cleaning up after dinner much faster. A product that aids a disabled user in undertaking everyday tasks more easily, for example an electronic wheelchair, can enable the user to move around more easily and be less isolated. Smartphones can allow busy families to keep in touch with each other quickly and easily.

> **Usability:** How easy a product is to use, how clear and obvious the functions are.

- A well-designed product can make a user's lifestyle easier or more time efficient.
- A badly designed product could lead to discomfort or injury.
- As a designer, observing how people interact with products is a good way to identify possible design solutions.

Ease of use and inclusivity

REVISED

- It is important to consider how easy a product is to use when designing.
- Inclusive design involves designing products in such a way that they are easy to use by as many people as possible.
- A designer should design with the widest possible user group in mind, excluding as few users as possible.
- Factors to consider when assessing inclusivity:
  - signage (can it understood without using words?)
  - lighting (should be adequate for the visually impaired)
  - visual contrast (clearly defined edges, clarity of design)
  - materials (texture, performance)
  - function (does what it needs to)
  - interface (how easy it is to use and understand)
  - intellectual access (can it be accessed by different intellects?)
  - emotional access (does it support emotional well-being?)
- An inclusive design should be:
  - inclusive – everyone can use it safely, easily and with dignity
  - responsive – taking account of what people say they need or want
  - flexible – different people can use it in different ways
  - convenient – everyone can use it without too much effort or separation
  - accommodating – all people regardless of their age, gender, mobility, ethnicity or circumstances should be able to use it
  - welcoming – no disabling barriers that may exclude some people
  - realistic – it should offer more than one solution to help balance everyone's needs, recognising that one solution may not work for all.

Ergonomic considerations and anthropometric data to support ease of use

- **Ergonomics** is the study of the relationship between people and the things they do, the objects they use, and their environment.
- A product that has been designed with ergonomics in mind is likely to be easier and more straightforward to use.
- **Anthropometrics** are human body measurements (for example, height, weight, finger length, hand-span).
- Anthropometric data can be used by designers to help to ensure a product is easy to use for all potential users. For example, anthropometric data can help designers to find the best height and width for a workstation.

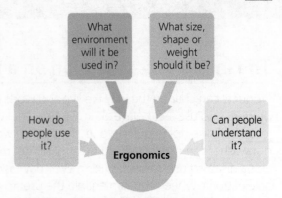

Figure 2.1 **What is ergonomics?**

Dimensions (mm)	Age range 5–9 Combined (percentiles)			Age range 13–18 Combined (percentiles)			Age range 19–65 Men (percentiles)			Women (percentiles)		
	5%	50%	95%	5%	50%	95%	5%	50%	95%	5%	50%	95%
1 Height	1,058	1,264	1,483	1,470	1,685	1,857	1,630	1,745	1,860	1,510	1,620	1,730
2 Eye level	895	1,055	1,180	1,456	1,570	1,740	1,520	1,640	1,760	1,410	1,515	1,620
3 Shoulder height	843	1,014	1,198	1,184	1,352	1,525	1,340	1,445	1,550	1,240	1,330	1,420
4 Elbow height	610	720	805	945	1,005	1,170	1,020	1,100	1,180	950	1,020	1,090
5 Hip height	496	619	754	734	855	990	850	935	1,020	750	820	890
6 Knuckle height (fist grip height)	375	480	565	690	720	815	700	765	830	670	720	770
7 Fingertip height	298	390	470	420	620	695	600	675	730	560	620	680
8 Vertical reach (standing position)	1,241	1,521	1,820	1,758	2,033	2,220	1,950	2,100	2,250	1,810	1,940	2,070
9 Forward grip reach (standing)	442	531	640	594	689	809	720	790	860	660	725	790

Figure 2.2 **Anthropometric data for standing**

> **Ergonomics:** The study of how we use and interact with a product or system.
>
> **Anthropometrics:** The study of the sizes of the human body.

> **Exam tip**
>
> Ensure you know the difference between ergonomics and anthropometrics. An exam question may ask you to explain how anthropometrics could be used to make a design ergonomically suitable for the user.

Aesthetic considerations

- As well as considering ease of use, designers also need to think about whether a design is aesthetically pleasing.
- **Aesthetics** are factors related to the appreciation of the beauty of a product. It is not only about what a product looks like; it may also consider how something sounds, feels, tastes or smells.
- The form, texture, scale, colour and symmetry of a product or system will have an influence on users' and stakeholders' opinions of it.
- Colour is an important consideration. Different colours can provoke different reactions – both positive and negative. Having an understanding of colour associations can help a designer to make the best choice for their product based on the response they are hoping to create.
- Designers often use the **golden ratio**. This is a mathematical ratio that is often found in nature and, when used, is thought to result in designs that are aesthetically pleasing and natural looking.

> **Aesthetics:** Factors concerned with the appreciation of beauty – this can include how something looks, sounds, feels, tastes and smells.
>
> **Golden ratio:** A common mathematical ratio found in nature that can be used to create pleasing, natural-looking designs.

Red		Aggressive, passion, strong and heavy, danger, socialism, heat
Blue		Comfort, loyalty, for boys, sea, sky, peace and tranquillity, conservativism, cold
Yellow		Caution, spring and brightness, joy, cowardice, sunlight
Green		Money, health, jealousy, greed, food, nature
Brown		Nature, aged and eccentric, rustic, soil and earth, heaviness
Orange		Warmth, excitement and energy, religion, fire, gaudiness
Pink		Soft, healthy, childlike and feminine, gratitude, sympathy
Purple		Royalty, sophistication and religion, creativity, wisdom
Black		Dramatic, classy and serious, modern, evil, mourning
Grey		Business, cold and distinctive, humility, neutrality
White		Clean, pure and simple, innocent, elegant, peace

Figure 2.3 Colour association

> **Typical mistake**
>
> Aesthetics isn't just about how something looks. It involves all the senses: sight, hearing, touch, taste and smell.

Now test yourself

1 Describe the meaning of the term 'usability'. [2 marks]
2 Describe the meaning of the term 'ergonomics'. [2 marks]
3 What is the use of human body measurements known as? [1 mark]
4 What term do we use when creating a design that has little or no barriers to the user? [1 mark]
5 What colour do we associate with royalty and sophistication? [1 mark]

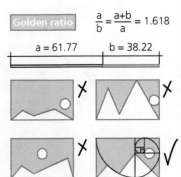

Golden ratio $\frac{a}{b} = \frac{a+b}{a} = 1.618$

a = 61.77   b = 38.22

Figure 2.4 The golden ratio

3 Opportunities and constraints that influence design and making

Analysing existing products with similar features can help a designer to understand user needs. It can also help them to consider how they may be able to improve upon an existing design with their own design solution.

When disassembling a product for analysis, you need to consider the following:

- function
- materials and components
- methods of construction and manufacture
- ergonomic and anthropometric considerations
- aesthetics, fashion and style
- how the product impacts the environment during its use
- ease of recycling
- how the product has been influenced by the work of past and present designers.

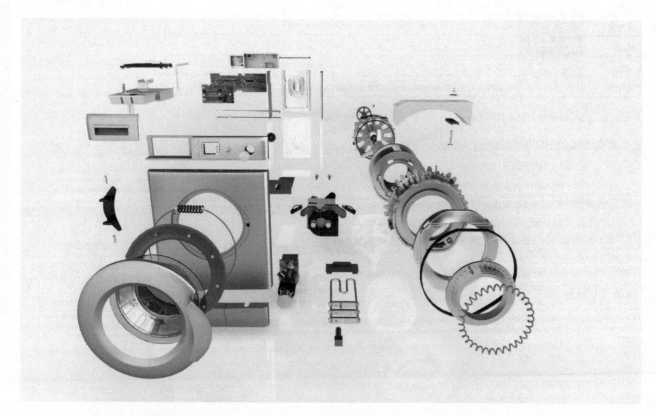

Figure 3.1 **Disassembling a product**

Materials, components and processes that have been used

REVISED

- Looking at particular materials, components and processes used to produce a product can help you to understand why they have been chosen. For example, which of the material's properties make it suitable for this purpose?
- By analysing materials, components and processes, you may be able to establish more suitable materials or different processes you could use to develop an improved product.

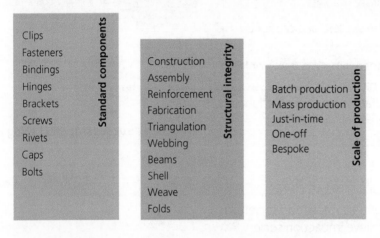

Standard components	**Structural integrity**	**Scale of production**
Clips	Construction	Batch production
Fasteners	Assembly	Mass production
Bindings	Reinforcement	Just-in-time
Hinges	Fabrication	One-off
Brackets	Triangulation	Bespoke
Screws	Webbing	
Rivets	Beams	
Caps	Shell	
Bolts	Weave	
	Folds	

Figure 3.2 What to look for during a product analysis

The influence of fashion, trends, taste and/or style

REVISED

- Designers are often influenced by distinctive design styles associated with a specific time period and try to emulate the characteristics of that style in their own work.
- Designers are also influenced by new trends and forecasts predicting future trends. These can change quickly.
- Technology has had a large influence on twenty-first century design trends.
- Increased consumer awareness about environmental issues and the need for sustainable design mean that eco design has become popular.
- In an increasingly globalised world, consumers are now more aware of the working conditions and cheap labour associated with manufacturing in some countries, and therefore **ethical** design has also become important.

> **Ethical:** Correct, good or honourable. Aim for an ethical approach when you are designing products.

| Victorian 1830s–90s | Art nouveau 1890–1905 | De Stijl 1914–31 | Bauhaus 1919–35 | Art deco 1925–39 | Streamlining 1930–50 | Organic design 1930–present | Scandinavian design 1935–present | Pop art 1960s | Minimalist design 1960s–present | Memphis 1980–90 |

Figure 3.3 Timeline of designers

The influence of marketing and branding

When developing new products, designers need to understand the requirements of their target market and whether there is likely to be a demand for their product.

The success of a product can be influenced by the **marketing** activity used to promote it.

- Marketing that appeals to the target market for a product can help to persuade people to buy it.
- Analysing the influence of marketing for existing products can help you to understand what might appeal to your target market.

Branding can also be a factor in determining the success of a product. When analysing existing products:

- Examine how the product uses branding – for example, a name or logo.
- Consider how important the product's brand has been to its success, or in encouraging people to buy the product. If a brand is popular or fashionable, it may be more successful than a comparable, unbranded product.

> **Marketing:** The business of promoting and selling a product; it can include advertising and promotion, and market research.

Figure 3.4 Successful marketing

The impact on society

When analysing products, you need to consider their impact on society in general. A successful product should have a positive impact on society. As humans, we interact and communicate with each other within our society. This is a very important part of our way of life, and part of human nature. Therefore, designers must consider the social impact of any new product and how it might affect wider society.

- People in different countries and cultures may have different feelings towards products based upon their own experiences.
- It is not only cultural differences – society is continually changing, and people's tastes and fashions reflect this.
- Sometimes a product can have a negative impact on society. For example, smartphones and tablets may allow us to communicate with people more easily regardless of where they are, but they can make us feel more isolated as we interact face to face less and less; they also make us more sedentary, contributing to increasing levels of obesity.

The impact on usability

When looking at an existing product:

- Identify its ergonomic features – how has it been designed to ensure a good 'fit' between it and its user so that it can be used easily?
- Consider how anthropometric data may have been used to ensure it is suitable for its users. What types of anthropometric data may have been considered?
- Think about ways in which you could improve usability and ergonomics.

The impact on the environment: lifecycle assessment

We live in a society influenced by consumerism and excessive consumption – a **throwaway society.** This can have a negative impact on the environment, as we both produce and dispose of more products.

When analysing existing products, it is also important to consider their impact on the environment.

- Think about the materials used – are they sustainable? Can they be recycled or reused at the end of the product's life?
- Consider energy consumption: does the product use a lot of energy to function? How much energy may have been used in the production process, or in the extraction of materials used to make the product?
- A **lifecycle assessment** can be used by a designer to analyse the impact a product has on the environment throughout its life. It considers the manufacture, use and disposal of the product.
- A designer can use a lifecycle assessment to consider how they can reduce the impact the product has on the environment.

> **Throwaway society:** A society influenced by consumerism and excessive consumption of products.
>
> **Lifecycle assessment:** The analysis of the impact of a product on the environment throughout the manufacture, use and disposal of that product.

> **Typical mistake**
>
> Remember that a lifecycle assessment will look at the impact a product has on the environment throughout its manufacture, use and disposal. Many candidates only consider the use and disposal of a product.

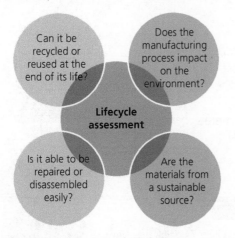

Figure 3.5 Lifecycle assessment

The work of past and present professionals and companies

- Many current and past designers and design companies have led the development of new technologies or the use of materials and particular aesthetics in products.
- Designers are often influenced by the work of these contemporary or past designers, companies or design movements.
- Designers are becoming more involved in collaborations with scientists and technologists to incorporate new technological developments into products.

> **Exam tip**
>
> Carry out some research into a designer of your choice from the past or present. Being able to demonstrate that you are aware of real designers by referring to a designer and their work may help to maximise your marks in the exam.

Now test yourself

TESTED

1. Explain the meaning of the term 'ethical'. [1 mark]
2. State four things to look for when undertaking a disassembly of a product. [4 marks]
3. What is the purpose of marketing? [1 mark]
4. What is the first part of a lifecycle assessment? [2 marks]

4 Developments in technology and their influence on design decisions and practice

Ethics

REVISED

- Designers need to consider how new technologies and the development of new products impact upon the **environment**. This means that the designer often needs to make ethical decisions.
- Ethical decisions are concerned with how companies:
  - manage waste
  - use resources.
- Ethical decisions usually put the needs of people and the environment ahead of profits.
- Examples of ethical decisions include:
  1 Understanding that the development of new products can put pressure on people to purchase the latest gadgets and that this will increase waste unless the old product is reused or recycled.
  2 Understanding that the mining of minerals and metals neededfor electronic gadgets can cause political problems around the world, as well as affecting landscapes and habitats.

> **Environment:** The natural and man-made world around us.

The environment

REVISED

- Waste generated by products can cause environmental issues, for example pollution.
- An obsolete product is one that is no longer of use, for example a 2016 calendar.
- Planned obsolescence means deliberately making a product out of date by stopping its supply or service support and by introducing a new model or version. For example, a washing machine is designed to require only minor repairs in the first two or three years but after four or five years vital parts will wear out and it will become uneconomical to repair.
- Sometimes planned obsolescence can be a positive thing. Using new products such as one-use medical syringes and disposal razors can avoid the spread of infection.
- Sustainability means meeting the needs of the present generation without compromising the ability of future generations to meet their own needs.
- Making products sustainable means considering the long-term effects that the product will have on the environment.
- The six Rs can be used as strategies to help develop sustainable products.

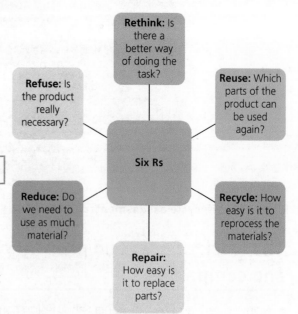

Rethink: Is there a better way of doing the task?

Refuse: Is the product really necessary?

Reuse: Which parts of the product can be used again?

Six Rs

Reduce: Do we need to use as much material?

Recycle: How easy is it to reprocess the materials?

Repair: How easy is it to replace parts?

Figure 4.1 **The six Rs**

Product enhancement

- A product enhancement is a change or upgrade to a product that increases its **capabilities**, for example a software upgrade.
- Meeting customer requirements increasingly requires upgrading of products with new technologies.
- If an upgrade is not **compatible** with older technologies it can cause older products to become obsolete.

> **Capabilities:** The abilities of a product to complete a task.
>
> **Compatible:** How a product fits or works with other products.

Typical mistake

When students are asked to think about how they can reduce the amount of material used to make a product, they tend to think about making the product smaller. In some cases this means that the product will not function as intended. For example, if you make a box for a toy smaller it is possible that the parts of the toy will not fit into the box. Remember: to successfully reduce the materials used to make a product it must still function as intended.

Exam tip

Remember: the pace of development of new products and technologies is rapid, so new products can become obsolete in a very short period of time. In an exam you may be asked to think about how a product can be enhanced or upgraded. For example, how could you enhance an ageing car to prolong its life?

Now test yourself

1 State two things that an ethical decision puts ahead of profit. [2 marks]
2 What is meant by the term an 'obsolete product'? [1 mark]
3 What is meant by the term 'planned obsolescence'? [1 mark]
4 State two strategies that can be used to design sustainable products. [2 marks]
5 What is meant by the term 'product enhancement'? [1 mark]

4 Developments in technology and their influence on design decisions and practice

5 The impact of new and emerging technologies when developing design solutions

New technologies are constantly emerging, so that existing technology either becomes outdated or needs to be **refined** and improved.

Industry and enterprise

REVISED

New and emerging technologies are having a large impact on industry and enterprise by reducing the number of staff required and increasing the skill levels of those staff that are needed. For example:

- **Artificial intelligence**, such as software that helps make online transactions, has many advantages but may result in increased unemployment.
- **Biometrics** are used in passports and allow people to use automated passport checks, reducing the need for passport control staff.
- **Virtual reality** is increasingly used for training; although it is expensive, it does allow training to take place that would otherwise be impossible, for example pilots practising emergency routines.
- **Drones** can carry out courier deliveries but technically skilled people will be needed to manage an efficient drone delivery network.

Circular economy

REVISED

A **circular economy** is an alternative to a traditional linear economy (make, use and dispose) in which:

- the resources are kept in use for as long as possible
- the maximum value is extracted from them while they are in use
- parts and materials are recovered and regenerated at the end of their life.

Cradle-to-cradle production

REVISED

In cradle-to-cradle production all materials and inputs and outputs are seen as either technical nutrients, which can be recycled or reused with no loss of quality, or biological nutrients, which can be **composted** or **consumed**.

People – lifestyle, culture and society

REVISED

All new technology will have an impact on our lives and, while it provides many benefits, there are also downsides. For example:

- Internet search engines bring the benefits of access to all sorts of information, but people can become addicted to using social networks and become victims of online fraud.
- Online shopping impacts on our environment by resulting in increased use of fuel for deliveries.

Exam tip

New and emerging technologies have many benefits but also some disadvantages. Remember to present a balanced picture if asked to answer a question about the impact of new and emerging technologies.

Refined: Minor changes to improve the product.

Circular economy: model in which resources are kept in use for as long as possible and the maximum value is extracted from them during and after their useful life.

Composted: Naturally broken down.

Consumed: Eaten.

Typical mistake

It is easy to assume that the demand for the latest products will continue to grow. This may not be the case however. Some consumers are now looking towards upgrading their old products rather than purchasing new ones.

The environment

- Designing products that last has become increasingly important in the twenty-first century, even though consumers always want the latest technology and designs.
- While some older products still function as intended, it becomes harder to access replacement parts and they do not always link to new technology (compatibility).
- Planned obsolescence (deliberately making a product out of date) has a significant impact on the environment. You can read more about planned obsolescence in Chapter 4.
- We can all contribute to saving energy and resources by:
  - making decisions on whether products and gadgets are actually necessary
  - using energy-efficient products, for example cars with good fuel consumption
  - using appliances efficiently, for example boiling just enough water in a kettle
  - choosing reusable products over disposable ones
  - buying products that can be upgraded or repaired
  - choosing products with minimal packaging, reducing waste.

Sustainability

- Sustainability refers to the concept of meeting the needs of the present generation without compromising the ability of future generations to meet their own needs.
- Designing with sustainability in mind means:
  - choosing non-toxic, sustainable or recycled materials
  - manufacturing and producing products using less energy
  - making products fuel and material efficient
  - producing products that are long lasting and better functioning so they do not need to be replaced
  - designing products that can easily be recycled when their use is finished (**disassembly**)
  - developing products that are profitable but also offer income to producers
  - considering the impact of a design on all stakeholders involved in its development.
- You can use the six Rs to assess sustainability – see Chapter 4.

> **Disassembly:** The taking apart of a product.

Now test yourself

1 Outline one example of how new and emerging technologies have resulted in a need for fewer staff. [2 marks]
2 Outline one example of how new and emerging technologies have resulted in a need for more highly trained staff. [2 marks]
3 State one potentially negative impact to the consumer of greater access to the internet for shopping. [1 mark]
4 Explain what is meant by the term 'circular economy'. [3 marks]
5 Explain what is meant by the term 'cradle-to-cradle' production. [2 marks]

6 Sources of energy

Generation of electricity

- There is a growing demand for energy throughout the world, especially from developing countries, and this has led to concerns over the environmental effects of using existing technology to meet these demands.
- The three fossil fuels – oil, natural gas and coal – are **finite**. This means that they are not **replenished** as they are used and are in danger of running out.
- Switching to alternative and renewable fuel sources means the non-renewable fuel sources will last longer.

> **Finite:** Limited in supply.
>
> **Replenished:** Quickly restored to previous levels.

How energy is stored and transferred

Energy can be stored in the following ways:

- **Kinetic energy:** the energy in a moving object. Dynamos and wind-up mechanisms transfer potential energy (stored until released) into kinetic energy.
- **Thermal energy:** energy that comes from heat. The Sun, radiators and fires give off thermal energy.
- **Chemical energy:** this is available in many different forms and is stored in fuels that we burn to release thermal energy. Batteries also use chemical energy.

> **Typical mistake**
>
> Remember that energy can neither be created nor destroyed; rather it transfers from one form to another. For example, chemical energy can be converted to kinetic energy in an explosion.

Non-renewable sources

- Non-renewable energy sources come out of the ground as liquids, gases and solids.
- Coal, crude oil and natural gas are all fossil fuels. They were formed from the buried remains of plants and animals that lived many millions of years ago. Over time they have created carbon-rich fuel sources.
- These fuels are burnt in power stations to generate heat (thermal energy), which is used to heat water and generate steam that turns turbines and generators (kinetic energy) to generate electricity (electrical energy).
- Uranium is not a fossil fuel but it is classified as a non-renewable energy source. It is mined and converted to a fuel used in power stations.
- Using fossil fuels to generate energy has a significant environmental impact. Burning fuels produces waste products, including gases such as sulphur dioxide, nitrogen oxides and other volatile organic compounds, which can have a harmful effect on the environment and contribute to global warming.

Table 6.1 **Non-renewable sources of energy**

Method	Description
Nuclear	Nuclear fission generates heat, which heats water to generate steam; steam turns turbines; turbines turn generators; electricity is distributed.
Gas/coal/oil	Fuel is burnt to generate heat, which heats water to generate steam; steam turns turbines; turbines turn generators; electricity is distributed.

Now test yourself answers and quick quizzes at **www.hoddereducation.co.uk/myrevisionnotes**

Renewable sources

- Renewable energy sources, such as sun light and wind, can be replenished naturally in a short period of time.
- Renewable energy sources are generally less harmful to the environment.

Table 6.2 Renewable sources of electricity

Method	Description
Hydroelectric	Dam is used to trap water, the water released turns turbines, turbines turn generators, electricity is distributed.
Wind	Blades are designed to catch wind, blades turn turbines using gears, turbines turn generators, electricity is distributed.
Solar photovoltaic	Photovoltaic cells convert light to electricity.
Tidal barrages	Barrage built across river estuary, turbines turn as tide enters (and when tide leaves), turbines turn generators, electricity is distributed.
Wave	Motion of waves forces air up cylinder to turn turbines, turbines turn generators, electricity is distributed.
Geothermal	Cold water is pumped underground through heated rocks, steam turns turbines, turbines turn generators, electricity is distributed.
Biomass	Fuel (wood, sugar cane, etc.) is burnt to generate heat, which heats water to generate steam; steam turns turbines; turbines turn generators; electricity is distributed.

Batteries

- A battery is a self-contained chemical power pack that can produce a limited amount of electrical energy wherever it is needed. A battery slowly converts chemical energy to electrical energy over a period of time.
- The disposal of used batteries has a large environmental impact.

> **Exam tip**
>
> In an exam you may be asked to talk about how recent developments in the design of a product have meant that it has a less harmful impact on the environment. A good example to use is a car, for example better fuel efficiency, lower emissions, use of recycled materials, development of electric cars.

Now test yourself

1 Name two fossil fuels. [2 marks]
2 What is meant by the term 'kinetic energy'? [1 mark]
3 What is meant by the term 'renewable energy source'? [2 marks]
4 What is meant by the term 'non-renewable energy source'? [2 marks]
5 Name one waste product that is produced when burning fossil fuels. [1 mark]

7 Environmental, social and economic influences

Designers must consider how new and emerging technologies impact on environmental, social and economic factors.

Environmental initiatives

REVISED

There are many environmental initiatives in place that aim to limit the impact of designing and making on the environment. These include:

- a **circular economy** where resources are used for as long as possible and then products and materials are reused and regenerated (see Chapter 5)
- the use of renewable sources of energy as alternatives to fossil fuels (see Chapter 6).

Fair trade

REVISED

- Fair trade is about establishing better prices, working conditions and terms of trade for farmers and workers.
- Many supermarkets and department stores stock fair trade goods, for example tea, coffee, chocolate and rice.

Figure 7.1 **Example of a fair trade product**

Social and ethical awareness

REVISED

- Products are made by real people and there are moral implications for us all when buying things.
- We need to consider the **social** and **ethical issues** involved in designing and making the products we buy. For example, the cotton industry has a huge ethical issue surrounding child labour and the welfare of workers.
- People can make a difference by being aware of social and ethical issues and making informed choices when designing or buying products. For example, buying clothes made from fair trade cotton helps low-paid cotton farmers around the world.

> **Social awareness:** How the designing, making and use of products impact upon people.
>
> **Ethical awareness:** Concerned with doing the right thing.

Typical mistake

You might think that the highest profit can only be made if the lowest cost materials are used. This is untrue. Some consumers are prepared to pay more for products made from more expensive materials that are likely to last longer.

Global sustainable development

- Global sustainable development, as defined by the Brundtland Commission, is development that meets the needs of the present generation without compromising the ability of future generations to meet their needs.
- When discussing sustainability we must consider that production processes require energy to process and source materials, which can cause pollution and affect climate change. This can damage people's health.
- The Paris Agreement of 2016 was the first Global Sustainable Development Agreement. It focuses on reducing greenhouse gases and emissions.
- The UK Sustainable Development Strategy recognises the need for a new, more environmentally sound approach to development in terms of energy production, transport and waste management.

> **Exam tip**
>
> Remember, the global market is changing. Many people are no longer only concerned with the cost and quality of products, but also about the working conditions and welfare of the people who made the products.

Now test yourself

1 Describe what is meant by the term 'fair trade'.   [2 marks]
2 Name two commonly available fair trade products.   [2 marks]
3 What is meant by the term 'social awareness'?   [1 mark]
4 What is meant by the term 'ethical awareness'?   [1 mark]
5 Name one agreement or strategy that has contributed to global sustainable development. [1 mark]

7 Environmental, social and economic influences

8 Communicating design solutions

2D and 3D sketches with notes

REVISED

- Sketches with notes are an easy and fast way to communicate ideas.
- Sketches can either be two dimensional (2D) or three dimensional (3D).
- Sketches can be produced with a pencil, pen or fine liner.
- Details and notes are added to sketches to help the designer create a more realistic view of the product and aid clarity.
- Sketching can be used effectively at the following stages in the designing and making process:
  - at the start of a design task to facilitate understanding and analysis of the problem and context
  - for initial design ideas
  - to explain a design concept
  - to help sell or promote the product.

Perspective drawing

- Perspective is the most realistic of all 3D drawing techniques because it works in the same way as our eyes; the further the object is away, the smaller it appears.
- In one-point perspective, all but the **horizontal** and **vertical** lines meet at a single vanishing point.
- In two-point perspective, all the lines except the vertical lines go to one of two vanishing points.
- In both one- and two-point perspective, the vanishing points are on the **eye line**.
- The object can be placed above or below the eye line depending on the view required. If an object is below the eye line, the top of the object will be seen. If an object is above the eye line, the underneath of the object will be seen.

> **Horizontal:** On its side or level with the horizon.
>
> **Vertical:** Upright.
>
> **Eye line:** The height of the eyes.

Figure 8.1 One-point perspective

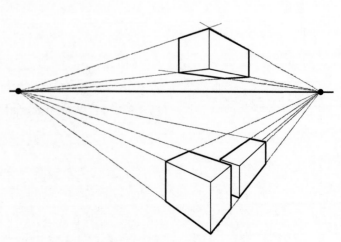

Figure 8.2 Two-point perspective

Isometric drawing

- Isometric drawing is a 3D drawing technique.
- In isometric drawing all the vertical lines remain vertical while the horizontal lines are drawn at 30 degrees.
- Ready-printed isometric grid paper can be placed underneath the paper to speed up the process.
- In isometric drawing a circle will appear as an **ellipse**.
- The crating/wire frame technique can be used to draw more complex shapes.

Ellipse: Also called an oval.

Distortion: A false view.

Dimensions: Measurements.

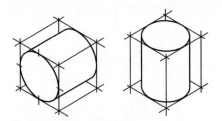

Figure 8.3 **Isometric drawing**

Oblique drawing

- Oblique drawing is a 3D drawing technique.
- The front or side view is drawn first and then depth added by projecting lines back at 45 degrees.
- The length of the 45 degree lines need to halved to avoid **distortion**.

Working drawings

- Working drawings include all the information needed to make a design, including:
  - **dimensions**
  - details of components
  - materials
  - assembly instructions.

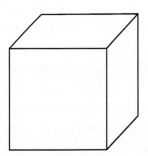

Figure 8.4 **Oblique drawing**

- Orthographic drawings are the most common method of producing working drawings. They can be produced by hand or using CAD.
- Orthographic drawing provides a set of 2D drawings of an object.
- The three most commonly drawn views in orthographic drawing are front view, end view and plan. The positions of these drawings on the paper is determined by the BSI (British Standards Institute).

Typical mistake

Features of a working drawing are often missing, for example the scale or dimensions.

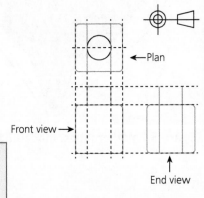

Front view →

←Plan

End view

Figure 8.5 **Orthographic drawing**

Sketch modelling

- Sketch models are simple physical models made from soft, low-cost, easy-to-work materials such as cardboard, Styrofoam, foam board or calico.
- Sketch models are used to explore or create initial ideas and can provide 2D and 3D models to physically test with users.

Exploded drawings

- Exploded drawings are 3D drawings that show how an object fits together.
- The components of the object are slightly separated, but all the components still line up with each other.

Mathematical modelling

- Mathematical modelling is the representation of a real situation, but rather than a physical model it uses mathematical concepts and language. For example, mathematical modelling and computerised simulation software can be used to test circuits and mechanical devices without the need to physically build them.
- The advantages of mathematical modelling are:
  - It is cost effective as no physical components are used.
  - It is quick and can speed up the production process.
  - It can predict stresses so that, if necessary, they can be strengthened before production begins.
  - It can assist with rapid prototyping.

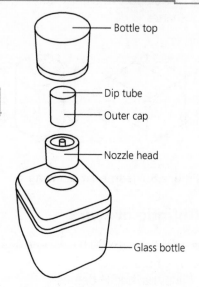

Figure 8.6 **Exploded drawing**

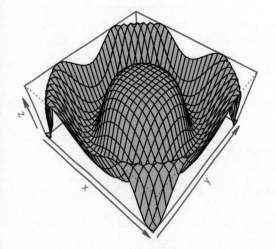

Figure 8.7 **Mathematical modelling surface mesh**

Flowcharts

- A flowchart is used to show a sequence of steps to make a product.
- Specific symbols are used for each stage in the process, for example start/stop, process and decision.
- The symbols are linked by arrows to show the direction.
- Flowcharts can be used to control quality by adding feedback loops to decision boxes.
- Schematic drawings are used to show the arrangement of components in electrical and mechanical systems.
- Schematic drawings are used to show relative points of interconnection rather than the actual position of the components.

> **Exam tip**
>
> You need to know how to use all of the drawing and sketching techniques in this chapter, as well as where to use them. For example, you would use an exploded drawing to show how the components or parts of a product fit together.

Figure 8.8 3D example of flowchart symbols

There are many different types of flowcharts; the most common symbols used in a flowchart are:

- Terminal – this signals the start or end of a process, and is usually shown as a diamond.
- Process – this explains an activity or a step in a process, and is shown by a rectangular box.
- Input or output – this is often shown by a rounded box.
- Decision – this is usually shown by a diamond.

Now test yourself

1 State two stages in the designing and making process where sketches can be used.   [2 marks]
2 Name two 3D drawing techniques.   [2 marks]
3 Name the three views that are commonly shown in orthographic drawings.   [3 marks]
4 State two advantages of mathematical modelling.   [2 marks]
5 Explain how a flowchart can be used to control quality.   [2 marks]

9 Sources of information and thinking when problem solving

When considering how to approach a design problem, designers often make use of a wide range of sources of information. This can be collected in various ways.

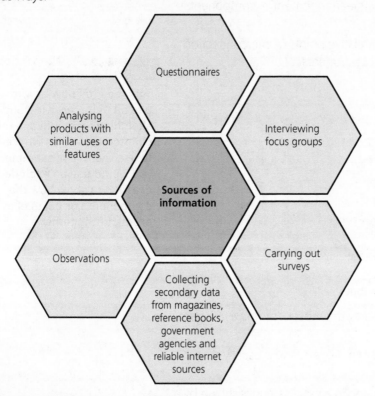

Figure 9.1 **Different sources of information**

Biomimicry

When designers look to nature and incorporate its approaches to problems into design solutions

Figure 9.2 **Biomimicry**

User-centred design

User-centred design (UCD) is a design strategy that focuses on understanding the intended user of a product and their experience. It considers the tasks they complete, the environment in which they live and work, and how they interact with a product.

- UCD is driven and refined by user evaluation and feedback.
- UCD is an iterative process and includes using focus groups, usability testing, participatory design, interviews and questionnaires to gather feedback.

Some methods commonly used in UCD are:

- focus groups, where an invited group of intended users share their thoughts, feelings, attitudes and ideas
- usability testing, which evaluates products by collecting data from people as they try out a product
- participatory design, which actively involves users in the design and decision-making process at every stage
- interviews, which are usually carried out with one participant at a time
- questionnaires, which ask users for their responses to a set of questions.

User-centred design (UCD): Sometimes called 'human-centred design', the user-centred design strategy aims to make products and systems useable. It focuses on the user interface and how the user interacts with and relates to the product.

Exam tip

In the exam there may be questions that focus on the users of a product: this may be a designing question or an analysis question. Try to consider all possible users and focus on user-centred design (UCD).

Systems thinking

Systems thinking is a design approach that considers the whole problem rather than focusing on one aspect, and which assesses all possible solutions.

- when using a systems approach to design, a designer looks at the whole experience associated with a product and not just the product itself – the product is part of a larger system of other products and systems.
- Considerations in systems thinking include the user's experience of the product, the marketing of the product, how it might be updated and maintained, and how it is disposed of, updated or exchanged at the end of its life.

Collaboration to gain specialist knowledge from across subject areas

- **Collaboration** – working with other designers, developers or stakeholders – during the design of a product allows designers to share ideas and to find solutions and improve a product.
- Designers increasingly work with specialists from other subject areas – for example, scientists or technologists – to benefit from expertise they may not possess themselves.

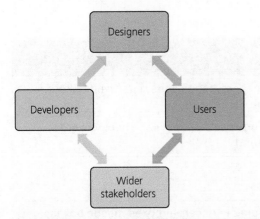

Figure 9.3 Effective collaborative working

> **Systems thinking:** The understanding of a product or component as part of a larger system of other products and systems. In the iterative design process, consideration of the role of all components and subsystems of the product or system, including the user experience and the marketing of the object being designed, ensures all aspects of the product are given the required attention to detail.

> **Collaboration:** Working with others for mutual benefit.

> **Typical mistake**
>
> Remember this phrase: 'A product is not just a product, it is actually a whole service'. Most candidates focus on the obvious when analysing a product or system. By looking at the possible wider issues, you will be able to access the higher marks, particularly in product analysis or extended writing questions.

Now test yourself

1 Give four sources of information when problem solving. [4 marks]
2 What does UCD stand for? [1 mark]
3 Explain the meaning of the term 'systems thinking'. [2 marks]
4 What term do we use to describe more than one person being involved in the iterative design process? [1 mark]

10 Categories of materials

Papers and boards

Papers and boards are used by designers for a range of purposes.

- Papers and boards come in a wide range of different thicknesses, sizes and types.
- The thickness, or 'weight', of paper is measured in grams per square metre (**gsm**).
- A weight of more than 170 gsm is classified as a board rather than a paper.
- The thickness of board is measured in microns. One **micron** is one-thousandth of a millimetre.

> **gsm:** Grams per square metre. Used to classify the weights of paper and card.
>
> **Micron:** One thousandth of a millimetre. Used to classify the thickness of paper and card.

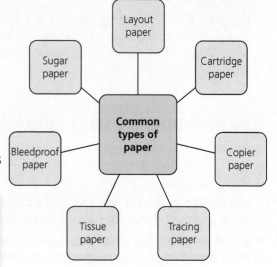

Figure 10.1 **Common types of paper**

Table 10.1 **Types of paper and board**

Paper or board	Description/properties	Uses
Paper	Available in a variety of colours, sizes and finishes	Newspapers, magazines, bus/train tickets, receipts, toilet paper
Card	• Slightly thicker than paper (180–300 gsm) • Available in a range of colours, sizes and finishes • Easy to fold, cut and print on	Greetings cards, paperback book covers, modelling
Cardboard	• Has a thickness greater than 300 microns • Available in different sizes and surface finishes • Inexpensive • Can be cut, folded and printed on easily	Packaging (e.g. cereal boxes, tissue boxes, sandwich packets), modelling, templates
Corrugated cardboard	• Strong but lightweight • Made from two layers of card with a fluted sheet in between • Thickness from 3 mm (3,000 microns) upwards • The fluted construction makes it very stiff and difficult to bend or fold	Packaging of fragile and delicate items; takeaway food packaging
Board sheets	• Rigid card • Thickness of 1.4 mm (1,400 microns) • Has a smooth surface • Available in different colours but white and black are most popular	Picture framing mounts; architectural modelling

Figure 10.2 **Corrugated cardboard can be used for takeaway pizza boxes**

Table 10.2 **Types of laminated board**

Laminated layers	Description/properties	Uses
Foam board	• A lightweight board consisting of polystyrene foam sandwiched between two pieces of thin card or paper • Has a smooth surface • Available in a range of colours • Different sheet sizes and thicknesses are available but 5 mm (5,000 microns) is the most common • Foam board is a very lightweight but rigid material • Easy to cut and fold	Modelling; point-of-sale displays
Styrofoam (expanded polystyrene foam)	• Available in a range of sizes and thicknesses • Blue in colour • Easy to cut, shape and sand to a smooth finish • Strong • Lightweight • Water resistant • Good heat-insulation properties	Wall insulation in caravans, boats and lorries; modelling; moulds for vacuum-formed or glass fibre products
Corriflute	• An extruded corrugated plastic sheet similar in structure and thickness to corrugated cardboard • Made from high-impact polypropylene resin • Available in a wide range of colours and sizes • Rigid • Lightweight • Waterproof • Easy to cut but can be difficult to fold, especially across the flutes	Estate agents' signs and signs outside shops; plastic containers; packaging; point-of-sale displays; modelling

Figure 10.3 **Styrofoam**

Figure 10.4 **Corriflute is used for outdoor signs**

Natural and manufactured timbers

Timber is the general name given to wooden planks and boards. Timber comes from trees that are cut down into logs and then sawn into planks. The three main types of timber are **hardwood**, **softwood** and **manufactured boards**.

Hardwood

- Hardwoods come from deciduous trees that shed their leaves every year.
- Hardwood trees grow slowly and can take hundreds of years to grow fully.
- Hardwood trees have thick trunks with branches at the top.
- Hardwoods have a close grain and tend to be denser, harder and heavier than softwoods.

Softwood

- Softwoods come from coniferous (evergreen) trees that have needles instead of leaves.
- Softwood trees keep their needles all year round.
- Softwood trees grow much faster than hardwood trees.
- Softwood trees grow tall and straight with lots of branches all the way up the trunk.
- Softwood has more knots than hardwood.

> **Hardwood:** Wood that comes from deciduous trees.
>
> **Softwood:** Wood that comes from coniferous (evergreen) trees.
>
> **Manufactured board:** Man-made boards made from recycled natural woods.

> **Exam tip**
>
> Not all hardwoods are necessarily hard. Some very soft, lightweight woods such as balsa are actually classed as hardwoods. Hardwood and softwood refers to the type of tree the wood comes from.

Softwood:
- mostly evergreen
- retain leaves all year round
- needle- or scale-like leaves
- bear cones.

Hardwood:
- mostly deciduous
- shed leaves each autumn
- typically flat leaves.

Figure 10.5 Hardwood and softwood trees

Figure 10.6 Knots in softwood are where the branches grew on the trunk of the tree

> **Exam tip**
>
> Manufactured boards are made from recycled woods so are environmentally friendly.

Manufactured boards

- Manufactured boards can be made from hardwood and softwood.
- Manufactured boards are made by gluing and compressing wood fibres or layers together.
- Manufactured boards are made in large sheets that are easy to work with.
- Manufactured boards are generally cheaper than 'real' or **natural wood**.

Table 10.3 **Examples of natural timbers and manufactured boards**

Hardwood	Softwood	Manufactured board
Oak	Scots pine	Medium-density fibreboard (MDF)
Mahogany	Spruce	Plywood
Teak	Cedar	Chipboard
Beech	Parana pine	Blockboard

Figure 10.7 **Plywood is made from thin layers of timber glued together**

Ferrous and non-ferrous metals

Metal comes from underground ores, which are extracted, processed and refined into metal bars, sheets and other forms ready for use. The two main types of metals are **ferrous metals** and **non-ferrous metals**.

Ferrous metals

- Ferrous metals contain iron.
- Ferrous metals are magnetic.
- Ferrous metals corrode quickly if not treated with a suitable surface finish.

Non-ferrous metals

- Non-ferrous metals do not contain iron.
- Non-ferrous metals are much more corrosion resistant than ferrous metals.
- Non-ferrous metals are generally more expensive than ferrous metals.

Figure 10.8 **Ferrous metals corrode quickly due to their iron content**

> **Natural woods:** Woods that come from trees.

> **Ferrous metals:** Metals that contain iron.
>
> **Non-ferrous metals:** Metals that do not contain iron.

Exam tip

Stainless steel is a ferrous metal but, unlike other ferrous metals, it is very resistant to rust due to the mixture of other elements it is made up of.

Typical mistake

All metals come from ore in the earth and so are non-renewable materials even though they can be recycled indefinitely.

Alloys

Alloys are metals made by combining two or more metals, and occasionally other elements, together to improve the properties in some way. The metals are carefully chosen and mixed to achieve specific properties such as improved hardness or strength, reducing the melting point or making the alloy more lightweight.

- Alloys are metals mixed or combined with other metals or substances.
- Alloys have specific properties derived from the metals they are made from.
- Alloys generally have one main ingredient with small quantities of other metals added.
- Alloys are generally cheaper than non-ferrous metals but more expensive than ferrous metals.

There are many different types of steel, which are made by mixing iron with small amounts of different metals and other elements. Although these are actually alloys, they are still classed as ferrous metals. For example, carbon steel is a mixture of iron with very small amounts (between 0. 5–2 per cent) of carbon.

Table 10.4 Examples of ferrous metals, non-ferrous metals and alloys

Ferrous metals	Non-ferrous metals	Alloys
Mild steel	Aluminium	Brass
Carbon steel	Copper	Pewter
Cast iron	Tin	Duralumin
Wrought iron	Zinc	Bronze

Thermo and thermosetting polymers

REVISED

Polymers are categorised as either thermo polymers or thermosetting polymers.

Thermo polymers

- Thermo polymers soften when they are heated and harden once they have cooled.
- Thermo polymers can be recycled as they can be reheated many times.
- When reheated, thermo polymers will return to their original shape. This is called plastic memory.

> **Alloys:** Metals mixed with other elements to improve their characteristics, for example hardness.

> **Exam tip**
>
> All metals, including alloys, can be recycled.

Figure 10.9 Brass is an alloy made from copper and zinc, giving it good resistance to wear and corrosion with an attractive appearance

> **Typical mistake**
>
> Copper will oxidise (and turn green) but not actually rust.

Table 10.5 Common thermo polymers

Common name	Properties/working characteristics	Uses
Polyethylene terephthalate (PET)	Clear, tough and shatter resistant, PET has good moisture and gas barrier properties	Soft drink bottles, mineral water bottles, fruit juice containers and cooking oil bottles
High-density polythene (HDPE)	Range of colours, hard, stiff, good chemical resistance, high impact	Milk crates, bottles, pipes, buckets, bowls
Polyvinyl chloride (PVC)	Stiff, hard, tough, good chemical and weather resistance, uPVC (unplasticized PVC) has strong resistance to chemicals and sunlight	Pipes, guttering, roofing sheets, window frames
Low-density polythene (LDPE) *blow moulding*	Range of colours, tough, flexible, good electrical insulator and chemical resistance	Washing-up liquid, detergent and squeezy bottles; bin liners and carrier bags
Polypropylene (PP)	Hard and lightweight, good chemical resistance, can be sterilised, good impact, easily welded together, resistance to work fatigue	Medical equipment, syringes, crates, string, rope, chair shells, containers with integral (built-in) hinges, kitchenware
Expanded polystyrene (EPS)	Lightweight, absorbs shock, good sound and heat insulator	Sound and heat insulation, protective packaging
Nylon	Hard, tough, resilient to wear, self-lubricating, resistant to chemicals and high temperatures	Gear wheels, bearings, curtain-rail fittings, clothing, combs, power-tool cases, hinges
Acrylic	Stiff, hard, clear, durable outdoors, easily machined and polished, good range of colours, excellent impact resistance (glass substitute), does scratch easily	Illuminated signs, aircraft canopies, car rear-light clusters, baths, Perspex sheet
Thermoplastic elastomers (TPE)	A combination of thermoplastics and elastomers; flexible and tough; after stretching and bending they will return to close to their original shape	Watch straps, scuba diving masks, remote control buttons
Acrylonitrile butadiene styrene (ABS) *Toys*	Tough, high-impact strength, lightweight, scratch resistant, chemical resistance, excellent appearance and finish	Kitchenware, safety helmets, car parts, telephones, food mixers, toys

Thermosetting polymers

- Thermosetting polymers undergo a chemical change when they set.
- Once thermosetting polymers have set they cannot be reheated or remoulded.

Table 10.6 Common thermosetting polymers

Common name	Properties/working characteristics	Uses
Urea-formaldehyde	Stiff, hard, brittle, heat resistance, good electrical insulator, range of colours	White electrical fittings, domestic appliance parts, wood glue
Melamine-formaldehyde	Stiff, hard, strong, range of colours, scratch and stain resistance, odourless	Tableware, decorative laminates for work surfaces, electrical insulation
Phenol-formaldehyde	Stiff, hard, strong, heat resistance	Dark electrical fittings, saucepan and kettle handles
Epoxy resin	Good chemical and wear resistance, heat resistance to 250°C, electrical insulator	Adhesives such as Araldite are used to bond different materials, for example wood, metal and porcelain
Polyester resin	Becomes tough when laminated with glass fibre, hard and strong, but brittle without reinforcement	GRP boats, chair shells, car bodies

Fibres and textiles

REVISED

Fibres are tiny hair-like structures that are spun (twisted) together to make yarns. These yarns are then woven or knitted together to create fabric.

Natural fibres

● Natural fibres are derived from both plants and animals.
● Examples of natural fibres include cotton, wool and silk.

Synthetic fibres

● **Synthetic fibres** are man-made.
● They come from many different sources including coal, oil, minerals and other petrochemicals.
● Examples of synthetic fibres include polyester, acrylic and nylon.
● These fibres are mostly **non-biodegradable** and are therefore not sustainable.
● Synthetic fibres are often manufactured to include properties such as flame resistance, crease resistance and stain resistance.

> **Exam tip**
>
> Know, understand and be able to give examples of the difference between natural and synthetic fibres, for example natural fibres – wool, silk, cotton; synthetic fibres – polyester, acrylic, nylon.

Mixed/blended fibres

● Fibres can be mixed or blended together to improve the quality, aesthetics, function or cost of a fabric.
● Fibres are mixed by adding yarns of different fibres together during the production process.
● An example of a mixed fibre is mixing elastane yarns with cotton yarns to give elasticity to a fabric.

Fibres: A thread or filament from which textiles are formed. They are tiny hair-like structures that are spun (twisted) together to make yarns. These yarns are then woven or knitted together to create fabric.

Synthetic fibres: Man-made fibres that come from a range of sources including coal, oil, minerals and other petrochemicals.

Non-biodegradable: Does not degrade or rot down into a harmless state.

Figure 10.10 **Fibres are spun together to make yarns, then woven or knitted together to create fabric**

- Blended fibres are blown together before they are spun into yarns.
- Polyester cotton is a common blend used to make shirts; it gives a fabric strength, breathability, absorbency and crease resistance, and is cheaper than a 100 per cent cotton shirt.

Woven fabrics

- Woven fabrics are produced on manual or automatic looms. A woven fabric has two sets of yarns: warp and weft yarns.
- Warp yarns run vertically.
- Weft yarns are woven horizontally under and over the weft yarns.

Figure 10.11 Polyester cotton is a popular blended fibre

Non-woven fabrics

- Non-woven fabrics are either bonded or felted.

Table 10.7 Non-woven fabrics

Bonded fabrics	Felted fabrics
Manufactured by applying pressure and heat or adhesives to bond the fibres together Usually used for single-use items as they lose strength and structure when wet	Manufactured by applying heat, moisture and friction to matt fibres together Usually use wool or acrylic fibres
Example: disposable textiles such as wet wipes, tea bags, surgical masks, dressings, nappies	Example: decorative appliqué, pool and snooker tables, cushioning and insulating

Knitted fabrics

- In knitted fabrics, yarns are looped together in rows of interlocking loops (stitches).
- Knitted fabrics are loose and flexible.
- The most commonly used knits are weft and warp.
- Weft knits include jersey and polyester fleece. Weft knits are used for socks, jumpers, scarves and hats.
- Tricot and knitted lace are examples of warp knits. Warp knits are used for swimwear, underwear and net curtains.

Figure 10.12 Disposable textiles such as surgical masks are made from bonded fabrics

Now test yourself

TESTED ☐

1 Name the two categories of polymers. [2 marks]
2 State one characteristic property of corrugated cardboard. [1 mark]
3 Name the most common thickness of foam board. [1 mark]
4 Describe the purpose of an alloy. [1 mark]
5 Name three manufactured boards. [3 marks]
6 Explain why fibres might be blended or mixed together. [3 marks]
7 Explain the main differences between hardwoods and softwoods. [4 marks]
8 Explain two advantages of manufactured boards compared to natural timbers. [2 marks]
9 State the three main differences between ferrous and non-ferrous metals. [3 marks]

11 Characteristic properties of materials

When choosing materials or components for a product, a designer needs to understand the characteristic properties of the material so that they are able to make an informed decision about whether those properties make it suitable for the intended purpose.

Density

REVISED

Density is the mass of a material (its weight) divided by its volume (size).

> **Density:** The mass of a material (its weight) divided by its volume (size).

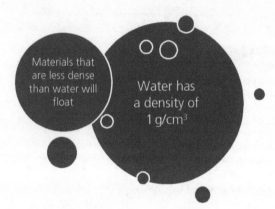

Materials that are less dense than water will float

Water has a density of 1 g/cm³

Figure 11.1 Density

Strength

REVISED

- The strength of a material is its ability to withstand forces that try to bend or break it.
- There are different kinds of strength, as shown in Figure 11.2.
- The strength of a material depends on the different forces that are applied to it and how well it can resist them. Some materials will be strong against one type of force but weak against another.
- Composite materials, timber and some polymers have high tensile strength but low compressive strength.
- Metals usually have tensile and compressive strength. The strength of textiles is sometimes referred to as **tenacity**.

> **Tenacity:** The strength of textiles and fabrics.

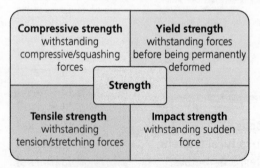

Compressive strength withstanding compressive/squashing forces

Yield strength withstanding forces before being permanently deformed

Strength

Tensile strength withstanding tension/stretching forces

Impact strength withstanding sudden force

Figure 11.2 Different types of strength testing

Hardness

REVISED

- The hardness of a material is how resistant it is to pressure from cutting, scratching or wear.
- Metals are harder than polymers and textiles.
- Hard materials are also often **brittle** (they will snap or break when bent).

> **Brittle:** A material that will shatter or break rather than bend and deform when forces are applied.
>
> **Durability:** The ability of a material to withstand wear, pressure or damage.

Durability

REVISED

- **Durability** is the ability of a material to withstand wear, pressure or damage.

Strength-to-weight ratio

REVISED

- Strength-to-weight ratio is a measure of a material's strength compared to its weight.
- Most polymers, carbon fibre, glass fibre and alloys (such as titanium) have high strength-to-weight ratios.
- Materials with high strength-to-weight ratios are often used in aircraft as they need to be strong but lightweight.

Stiffness

REVISED

- The stiffness of a material is its ability to resist being deformed when a force is applied to it (its rigidity).
- Ceramics, glass and steel are rigid.
- Polymers (for example foam and rubber) have poor stiffness.

Figure 11.3 **Steel is strong, hard and stiff – properties that make it suitable for engineering and construction**

Elasticity

- Elasticity is the ability of a material to return to its original shape when forces are applied to it that will make it bend or flex out of shape.
- Different materials have different levels of elasticity.
- Materials with low elasticity will deform very little before breaking.
- A material with high elasticity will return to its initial shape and size when forces are removed.

Impact resistance

- Impact resistance is the ability of a material to withstand a force or shock applied to it.
- Soft metals, such as mild steel, and polymers, such as rubber and nylon, have good impact resistance.

Plasticity

- **Plasticity** is the ability of a material to permanently change in shape when force is applied to it.
- Metals such as sheet steel are shaped by hammering or bending.

> **Plasticity:** The ability of a material to permanently change in shape when force is applied to it.

Corrosive resistance to chemicals and weather

- **Corrosive** resistance is how susceptible a material is to degradation from oxygen, moisture and other chemicals.
- Rust is a form of corrosion that affects ferrous metals when they are exposed to oxygen and moisture.
- Oak and dense hardwoods are resistant to corrosion.
- Softwood will rot (corrode) when exposed to the elements and left untreated.
- Many polymers degrade gradually over time if exposed to ultra violet light, oxygen or chemicals such as chlorine. Car and bicycle tyres, for example, will crack and split if left in the sun.

> **Corrosion:** The degradation of a material from elements such as oxygen, moisture and other chemicals.

Flammability

- Flammability is the ability of a substance to burn or ignite.
- Most timbers, polymers, fabrics, papers and boards are flammable.
- Metals are not flammable.

Figure 11.4

Absorbency

- Absorbency is the ability of a material to absorb moisture.
- Natural fabrics (for example, cotton, linen and wool), cardboard and foam have good absorbency.

Thermal conductivity

- Materials with good thermal **conductivity** allow heat to be transferred through them easily (and they will heat up).
- Metals are usually good conductors of heat.
- Materials with good thermal resistance will not let heat be transferred through them easily – they are good insulators.
- **Porous** materials such as Styrofoam and softwoods are good insulators.
- Thermal fabrics (for example, acrylic, wool and viscose) also have good **insulation** properties. They are used to keep the wearer warm in clothing, but also to keep things cool, such as in cool bags.

> **Conductivity:** How easily electricity, heat or sound is transmitted through a material.
>
> **Porous:** A material that has many tiny holes that allow moisture to seep through.
>
> **Insulation:** A material that prevents heat, electricity or sound from escaping.

Figure 11.5 Wool has good insulating properties

Electrical conductivity

- Electrical conductivity refers to how easy it is for electricity to flow through a material.
- Metals are good electrical conductors.
- Wood and rubber are poor electrical conductors.

Magnetic properties

A material with magnetic properties will emit forces that attract or repulse other materials.

Exam tip

Make sure you understand the characteristics of a range of materials. In the exam you will be expected to demonstrate that you can identify characteristics of common materials.
This may be describing the characteristics of a material or identifying and naming the material that has certain characteristics.

Typical mistake

Candidates often use words such as 'strong' or 'sturdy' when describing a material characteristic. Try to add a bit more detail or a more specific characteristic, such as good strength-to-weight ratio, dense, rigid, impact resistant.

Now test yourself

1 What does 'compressive strength' mean? [2 marks]
2 When might the term 'tenacity' be used? [2 marks]
3 Ferrous metals are corrosive, true or false? [1 mark]
4 Give one characteristic of an insulating material. [1 mark]
5 Name one fabric that has good insulating properties. [1 mark]

...ed movement

There are four basic types of **motion**.

Motion: Movement.

Rotary motion

- Rotary motion follows the path of a circle, for example the rotation of wheels or an electric motor shaft.
- Rotary motion can be measured by counting the number of revolutions in a set period of time, for example revolutions per minute (RPM).

Figure 12.1 **Wheels are an example of rotary motion**

Linear motion

- Linear motion is in a straight line, for example a vehicle travelling in a straight line or items on a conveyor belt.
- The speed at which objects travel in a straight line is measured by dividing the distance they travel by the time taken (speed = distance ÷ time).
- Speed is often measured in metres per second (m/s) or kilometres per hour (km/h).

Figure 12.2 **Items on a conveyor belt move with linear motion**

Oscillating motion

- Oscillating motion is similar to circular motion, but the motion moves back and forth in a circular path. An example is the head of an electric toothbrush.
- The rate of oscillating motion is measured in oscillations per second or per minute.

Figure 12.3 **An electric toothbrush head uses oscillating motion**

Reciprocating motion

- Reciprocating motion is back and forth in a straight line. An example is the blade on an electric jigsaw.
- The rate of reciprocating motion is also measured in oscillations per second or per minute.

How mechanical systems produce different types of motion

- A mechanism is a series of parts that work together to control forces and motion in a desired way. Controlled motion is essential in many engineered products.
- A force is a push, a pull or a twist. Forces are measured in Newtons (N).
- Mechanisms can be used to change:
  - one kind of motion to a different kind, for example linear motion to rotary motion.
  - the speed and direction of the motion.
  - the size and direction of forces.

Figure 12.4 **A jigsaw blade moves with reciprocating motion**

The effect of forces on the ease of movement

- Mechanisms that control and change motion have an input and an output. For example, on a pair of scissors the input motion is the movement on the handles and the output motion is the movement of the blades.
- Mechanisms can either:
  - **reduce** the distance moved but increase the force being exerted, or
  - **increase** the distance moved but reduce the force being exerted.
- Mechanisms can be used to increase or **amplify** a force but the distance moved will decrease. For example, in a car jack the distance moved by the handle is far greater than the distance the car is raised.

> **Reduce:** make smaller.
>
> **Increase or amplify:** make larger.
>
> **Fulcrum:** the point around which a bar or beam rotates

Levers

- The simplest example of a mechanism that controls and changes motion is a lever.
- A lever is a rigid bar that pivots on a **fulcrum**. The input force is called the **effort** and the output force is called the **load**.

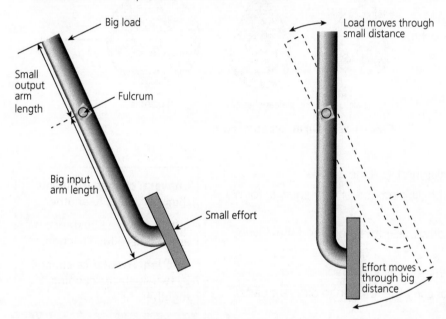

Figure 12.5 A brake pedal is a lever

- In Figure 12. 5 the position of the fulcrum means the effort moves through a larger distance than the load. This means that the brake pedal has effectively amplified the force applied by the driver's foot.
- The distance between the fulcrum and load is called arm length. The larger the arm length, the larger the distance the force must move, therefore, the smaller the force.

How different mechanical devices are used to change the magnitude and direction of motion or forces

- Mechanisms that control and change rotary motion also have an input and output, but the force is referred to as **torque**.
- Torque is a turning or twisting force. Rotary mechanical systems can either:
  ○ reduce rotary speed but increase torque, or
  ○ increase rotary speed but reduce torque.

Cams

- A cam and follower is a mechanism to **convert** rotary motion into reciprocating motion.
- A cam is a specially shaped wheel that the follower rests on. As the cam rotates, the follower moves up and down.
- The profile (shape) of the cam determines the motion of the follower throughout one rotation cycle.

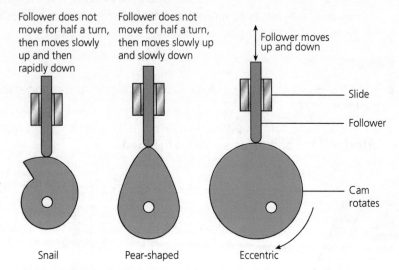

Follower does not move for half a turn, then moves slowly up and then rapidly down

Follower does not move for half a turn, then moves slowly up and slowly down

Follower moves up and down

Slide

Follower

Cam rotates

Snail   Pear-shaped   Eccentric

Figure 12.6 Three types of cam

Gears

- A spur gear is a wheel with teeth around its edge.
- The teeth of a spur gear mesh with (link into) other gear spurs to form a gear train.
- A simple gear train consists of a driver gear (the input) and the driven gear (the output).
- If the driver gear is the smaller gear, it is called a pinion.
- Gears are mounted on shafts that carry the rotation to different parts of the mechanism.
- If the two gears in a simple gear train are different sizes, they will rotate at different speeds. The simple rules are:
  ○ The smaller gear will rotate faster than the larger gear.
  ○ The gears will rotate in opposite directions.
- A gearbox is a mechanical system that contains several simple gear trains working together to achieve a very large speed reduction.
- Gear systems usually require some form of **lubrication** to reduce **friction** and to prevent the teeth wearing away, for example oil or grease.

> **Convert:** change from one thing, or state, to another.
>
> **Lubrication:** a substance applied to reduce friction.
>
> **Friction:** resistance caused by two surfaces rubbing together.

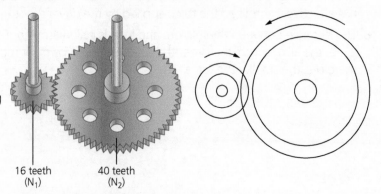

16 teeth (N_1)   40 teeth (N_2)

Figure 12.7 A simple gear train

Pulleys and belts

- Pulleys and belts transfer rotary motion between two shafts but are different from gear trains in the following ways:
  - The input and output shafts can be separated by a greater distance.
  - The input and output shafts rotate in the same direction.
- A pulley and belt drive can provide a reduction or increase in speed.

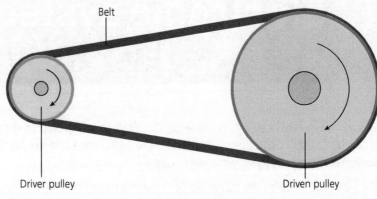

Figure 12.8 Transferring rotary motion with a pulley and a belt

Levers and linkages

- A linkage is a component used to direct forces and movement to where they are needed.
- A linkage can be used to:
  - change the direction of motion, or
  - **convert** between different types of motion.

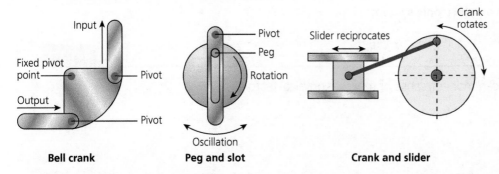

Figure 12.9 Examples of linkages

12 Controlled movement

Exam tip

Students are often confused about the names and terms associated with controlling movement. The table below will help you remember.

Motion	Lever	Change (magnitude or direction)
Rotary	Pivot	Cams
Linear	Effort	Gears
Oscillating	Fulcrum	Pulleys and belts
Reciprocating		Levers and linkages

Typical mistake

Students often label the direction of movement incorrectly on gear trains and belts and pulleys. Remember:
- Gear trains – gears move in opposite directions.
- Belts and pulleys – pulleys move in the same direction.

Now test yourself

TESTED ☐

1. Name four types of motion. [4 marks]
2. What unit is force measured in? [1 mark]
3. Sketch a cam and follower. [2 marks]
4. What unit are two gear wheels that mesh together called? [1 mark]
5. Sketch and label a pulley and belt system. [2 marks]

13 Electronic systems

Electronic systems are circuits made up of different components that are designed to carry out a specific function. We use a wide range of different electronic devices and systems every day.

All electronic systems are made up of three main stages:

- input
- process
- output.

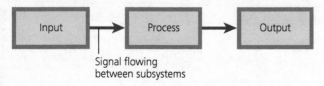

Figure 13.1 **The stages in an electronic system**

Inputs

- Inputs are sensors that can monitor, sense or measure physical factors such as:
  - temperature
  - pressure
  - light levels
  - weight
  - sound
  - movement
  - magnetic fields.
- Sensors produce an electrical **signal** in response to what they are sensing.
- This signal is sent to the processing part of the system.
- On many systems a push button is the sensor that sends a signal when it is pressed.

> **Signal:** An electronic voltage that is used to represent information.

Input devices

Push buttons and switches

- Push buttons and switches are the simplest **input devices**.
- They have only two signals – on and off.
- When not pressed, they send an 'off' signal.
- The 'off' signal means there is no voltage being sent.
- When pressed or switched on, they send an 'on' signal.
- The 'on' signal means all the voltage in the circuit is being sent.

> **Input device:** A device that senses something, for example temperature.

Table 13.1 Types of switches

Push switches	• Produce a signal when push button is pressed • Can be momentary (turns off when released) or latching (stays on until pressed again) • Available in many different shapes, sizes and colours
Rocker switches Toggle switches Slider switches	• Produce a signal when turned on • Used in many different household electrical appliances and other applications • Available in many different shapes, sizes and colours
Micro-switches	• Can be wired to send a signal when pressed or not pressed • Very small so can be 'hidden' in confined spaces • Wide range of uses such as game controllers, motorcycle brake levers and electric gates

Movement sensors

There are various different types of movement sensors.

- Tilt switches are switches that turn on when they are moved by a set amount.
- Tilt switches produce a signal when upright.
- They produce no signal when tilted by a certain amount.
- They are used for sensing if an item is falling over, for example a room heater or motorcycle.

Infrared (IR) sensors

These are another common type of movement sensor.

- An IR produces a signal when it senses warm objects within a certain range.
- A PIR (passive infrared) sensor produces a signal when it senses moving warm objects.
- Infrared movement sensors are often used in automatic lights, taps and hand dryers in public toilets.
- Passive infrared sensors are used in domestic and commercial burglar alarm systems.

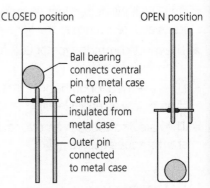

CLOSED position   OPEN position

Ball bearing connects central pin to metal case

Central pin insulated from metal case

Outer pin connected to metal case

Figure 13.2 Tilt switches use liquid mercury to turn current on and off

Figure 13.3 PIR sensor in a domestic burglar alarm

Magnetic field sensors

- Reed switches are turned on and off by a magnetic field.
- Reed switches produce no signal when a magnetic field is detected.
- They produce a signal when the magnetic field is removed.
- Used in house alarms to detect when a door or window has been opened (a magnet is mounted on the door next to the switch).

Temperature sensors

A common sensor used to sense temperature is a thermistor.

- A thermistor is a type of resistor that changes its level of resistance as the temperature rises.
- There are two main types of thermistor:
  - NTC (negative temperature coefficient) thermistors decrease their resistance (allow more current to pass through them) as the temperature increases.
  - PTC (positive temperature coefficient) thermistors increase their resistance (allow less current to pass through them) as temperature rises.

Light sensors

- An LDR (light-dependent resistor) is another type of resistor that changes its level of resistance as the light level increases.
- The most common type of LDRs decrease their resistance as the light intensity increases.
- LDRs are used for outdoor security lights that turn on when light levels drop below a certain level.
- They are also used in cameras to measure the light level and adjust the shutter speed accordingly.

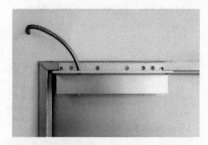

Figure 13.4 A magnetic reed switch used to detect when a door is opened

Figure 13.5 A thermistor

Figure 13.6 An LDR

Outputs

- **Output devices** produce audible or visual indicators such as light or sound.
- They can also produce movement, power or activate something such as a motor, solenoid or hydraulic valve.
- There are many different types of output device. The three most common are:
  - lights (visual indicators)
  - speakers and buzzers (audible indicators)
  - motors (produce movement).

Output device: A device that produces audible or visual indicators, or movement.

Typical mistake

Microphones are input devices, not output devices – they sense sound and convert it into an electrical signal.

Light outputs

In electronics, there are two main types of component that produce light outputs: lights bulbs or lamps, and light-emitting diodes (LEDs).

Light bulbs or lamps

● Light bulbs produce light by electric current that passes through a filament making it glow white-hot.
● Light bulbs have no polarity (can be connected to positive and negative any way around).
● Light bulbs need connecting to a socket which can be a screw-thread or two metal pins.
● Light bulbs use more power and are less efficient than LEDs.

Light-emitting diodes (LEDs)

● LEDs produce light by a flow of electrons across a band gap in a semiconductor.
● LEDs do have a polarity and will only work if connected the right way round.
● LEDs are much smaller than light bulbs.
● LEDs use less power and last much longer than light bulbs.
● LEDs are available in a wide variety of shapes, sizes and colours, including infrared and ultraviolet.

Due to their high efficiency, compact size and range of colours, LEDs have replaced light bulbs in most lighting applications. They are widely used in domestic lighting applications, household electrical products, street lighting, vehicle lights and children's toys.

Figure 13.7 A two-pin and screw fitting small light bulb

> **Typical mistake**
>
> If you use a touchscreen on a tablet or phone as an example, you should state that is an input *and* output device.

Speakers and buzzers

Speakers and buzzers are the most common audible output devices.

Speakers

● Speakers convert electrical audio signals from a microprocessor into an audible sound.
● Speakers can produce a wide variety of different sounds, from simple clicks to complex tunes, speech and music.
● Speakers require a waveform signal to work, not just a power connection.
● Active speakers contain an inbuilt amplifier to increase the strength of the signal.
● Wireless speakers receive audio signals using radio frequency (RF) waves rather than wires.

Figure 13.8 LEDs

Buzzers

- Buzzers convert an electric signal into an audible sound.
- Buzzers do not require a waveform signal; they just need to be connected to power.
- Buzzers cannot produce music or speech; they just produce a tone.
- Some buzzers can change tone depending on the strength of the signal (voltage level) sent to the buzzer.
- Loud buzzers are called sirens and are used in alarm and warning systems.

Motors

Electric motors are the most common type of output device used to produce motion. The rotary motion can then be converted into other types of motion to perform specific functions, such as opening locks, doors and gates, or powering machinery of some kind.

- Motors produce high-speed rotary motion when they receive power.
- Motors often use a gearbox to reduce and control speed.
- By changing the signal sent from the microprocessor, motors can be made to rotate forwards, backwards and vary their speed.
- Motors are used to create vibration by adding offset weight to the motor spindle.

> **Typical mistake**
>
> Remember that electronic signals sent to headphones and speakers carry no voltage.

Figure 13.9 Sirens are used in school fire alarms

Figure 13.10 A simple motor

Programmable components (microcontrollers)

- A **microcontroller** is a small computer **programmed** to carry out a specific task.
- Programmable microcontrollers take information received from input devices and process it into electrical signals which are then sent to different output devices.
- The microcontroller contains a program that is downloaded to the microprocessor's memory.
- The program is a set of instructions that tell the microcontroller what to do with the information it receives from the input devices.
- Microcontrollers are used in **embedded** applications, which means the program cannot be changed by the user.
- Microcontrollers are used in a wide range of devices and appliances, such as household products, power tools, toys and engine management systems.
- In many products only a fraction of the microcontroller's full potential is used.

> **Microcontroller:** A programmable electronic component.
>
> **Program:** A set of instructions loaded on to a microcontroller.
>
> **Embedded:** A program in an appliance that cannot be altered by the user, for example a washing machine.

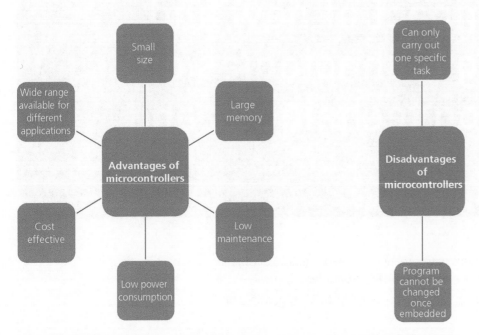

Figure 13.11 Advantages and disadvantages of microcontrollers

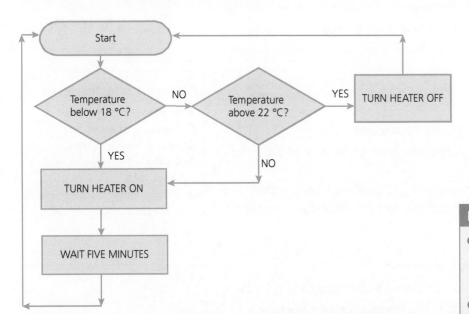

Figure 13.12 A simple electronic system diagram for a room heater

- Different types of microcontroller are available for use in different applications including GCSE projects.
- Generic pre-programmed microcontrollers designed for a specific purpose are available and can be modified by a programmer to enhance their functions. For example, an automatic gate that opens when it senses someone approaching could be modified to also turn on a light if they approach when it's dark, or to only open during certain times of the day.

Now test yourself

1. Name three different types of input device. [3 marks]
2. Name three different output devices. [3 marks]
3. Describe the role of a microcontroller in an electronic system. [3 marks]
4. Explain the effect of an embedded program in an appliance. [2 marks]
5. Explain the difference between a speaker and a buzzer. [2 marks]

[handwritten note: Are these the same Question ??]

14 The impact of new and emerging technologies on production techniques and systems

- Global communication systems have greatly increased the flow of digital information in text, visual and audio format. For example:
  - Large electronic documents can be transferred instantly regardless of distance.
  - The internet has enabled wider opportunities for research.
  - Marketing, advertising and sales opportunities have increased greatly.
- The traditional manufacturing industry is undergoing a digital transformation, including:
  - artificially intelligent robots
  - **autonomous** drones
  - 3D printing
  - nanotechnology (technology on a microscopic scale)
  - cloud computing (a network of online servers that store and manage data)
  - the **Internet of Things** (where electronic devices connect within the existing internet infrastructure, to send and receive data without human intervention).
- The digital transformation is:
  - reducing manufacturing production costs
  - improving manufacturing precision, speed, efficiency and flexibility.
- Scientific developments have led to the introduction of many new materials and composites with improved properties; for example, self-healing polymers and bio-materials that can be grown into products.
- New technologies bring many benefits but may also result in some negatives, for example increased unemployment due to jobs that were traditionally performed by human beings now being completed by robots.

> **Autonomous:** Acting alone.
>
> **Internet of Things:** the connection of everyday devices to the internet, allowing them to send and receive data.

Economies of scale

REVISED

- Economies of scale are the cost advantages that a manufacturer gains as a result of the size, output or scale of their production. Traditionally for larger-scale production, costs per unit are lower because:
  - fixed costs, such as machinery, are spread over more units
  - bulk buying reduces the costs of materials
  - marketing and advertising costs are spread over more units.

- In addition, larger-scale production allows:
  - Workers to specialise in producing a specific part of a product. This requires less training and can result in more efficient production.
  - Components can be produced in factories around the world, with the final assembly closer to home. Although this can increase shipping costs, it does allow companies to access benefits such as cheaper labour and lower duty charges.
- Newer technologies mean that some manufacturers can produce products in smaller quantities but still benefit from the reduction in costs associated with higher levels of production, for example by using 3D printing.
- Companies that produce bespoke products are able to:
  - meet the needs of individual clients more easily
  - react to fashions and trends
  - reduce the costs of storing final assembled products.

> **Exam tip**
>
> In an exam make sure that you are able to explain both the advantages and disadvantages of large-scale production. For example, fixed costs, such as machinery, will be lower as they are spread over more units, but you may be less able to meet individual client needs.

Disruptive technologies

REVISED

- A disruptive technology is one that displaces an established technology and shakes up the industry, or a ground-breaking product that creates a completely new industry. Examples of disruptive technology include:
  - **Additive manufacturing** – slicing up 3D products into a series of layers that are then sent for rapid prototyping; for example stereo lithography, laser sintering or 3D printing.
  - **Advanced robotics** – industrial robots are used extensively in industries such as automotive manufacturing for welding, material handling, painting and assembly.
  - **The Internet of Things** – allows electronic devices connected to each other to communicate with one another over the internet without human intervention. This can result in reduced downtime and fewer problems and defects with final products.
  - **Virtual reality** – VR is used to test products in a **virtual world** before production.
- The development of disruptive technology has had the following benefits:
  - increased flexibility to make complex products
  - increased use of digital modelling and simulation tools
  - improved productivity and quality
  - improved use of material resources and reduced energy consumption.

> **Typical mistake**
>
> Global communication systems, such as the internet, are only seen as a benefit to the manufacturing industry. Make sure that you are aware of the possible disadvantages, for example allowing consumers to purchase products from anywhere in the world and therefore increasing transportation costs.

> **Virtual world:** Not real, imaginary.

Now test yourself

TESTED

1 Name one benefit of the digital transformation of the manufacturing industry. [1 mark]
2 What is meant by the term 'economies of scale'? [1 mark]
3 State two benefits of bespoke production. [2 marks]
4 What is meant by the term 'disruptive technology'? [2 marks]
5 Name two examples of disruptive technology. [2 marks]

15 Papers and boards

Physical and working properties REVISED

Paper

Paper is a thin, versatile material that is easy to work with.

- Paper made from 'new' wood fibres is called **virgin fibre paper**.
- **Recycled paper** made from refined waste paper is now common.
- Paper is available in many different types, sizes, finishes and weights.

> **Virgin fibre paper:** Paper made from 'new', unused wood fibres.
>
> **Recycled paper:** Paper made from used paper products.
>
> **Lightweight:** Weighs very little.
>
> **Fluted:** A shape or object that has grooves or ridges.
>
> **Rigid:** Difficult to bend.

Figure 15.1 **Paper is a common and versatile material**

Table 15.1 **The properties, characteristics and uses of common papers**

Common name	Weight (gsm)	Properties/working characteristics	Uses
Layout paper	50	Bright white, smooth, lightweight (thin) so slightly transparent and inexpensive	Sketching and developing design ideas; tracing parts of designs
Copier paper	80	Bright white, smooth, medium weight, widely available	Printing and photocopying
Cartridge paper	80–140	Textured surface with creamy colour, thicker and more expensive than layout or copier paper	Drawing with pencils, crayons, pastels, watercolour paints, inks and gouache
Bleed-proof paper	80–140	Bright white, smooth surface, stops marker 'bleed'	Drawing with marker pens
Sugar paper	100	Available in wide range of colours, sold in larger sheet sizes, inexpensive, rough surface	Mounting and display work

Boards

A weight of more than 170 gsm is classified as a board rather than a paper.

Table 15.2 The properties, characteristics and uses of common boards

Common name	Thickness (microns)	Properties/working characteristics	Uses
Card	180–300	Available in a wide range of colours, sizes and finishes; easy to fold, cut and print on	Greetings cards, paperback book covers, simple modelling
Cardboard	300 upwards	Available in a wide range of sizes and finishes; inexpensive; easy to fold, cut and print on	General retail packaging for food, toys, etc.; design modelling
Corrugated cardboard	3,000 upwards	Lightweight yet strong; **fluted** construction makes it difficult to fold; can absorb knocks and bumps; good heat insulator	Pizza boxes, shoeboxes, larger product packaging, e.g. electrical goods; packaging for delicate or fragile items
Mounting board	1,400	Smooth; **rigid**; good fade resistance; available in different colours	Borders and mounts for picture frames

Laminated layers

Table 15.3 The properties, characteristics and uses of common laminated layers

Common name	Properties/working characteristics	Uses
Foam board	Smooth, rigid, very lightweight and easy to cut and fold, available in a range of colours, sheet sizes and thicknesses	Point-of-sale displays, ceiling-hung signs in supermarkets, architectural modelling
Styrofoam	Light blue colour; easy to cut, sand and shape; strong and lightweight; water resistant; good heat and sound insulator; available in a wide range of sizes and thicknesses	3D moulds for vacuum forming and GRP, wall insulation in caravans, boats, etc.
Corriflute	Made from a high-impact polypropylene resin, wide range of colours, waterproof, easy to cut but can be difficult to fold, rigid, lightweight	Outdoor signs, containers, packaging and modelling
PVC foam sheet	Lightweight, good insulation properties, easy to print on, easy to cut and join to other materials, water resistant, available in a range of different colours	Outside displays and signs

Factors that influence selection

When selecting materials and components, as well as considering their physical properties and working characteristics, a designer needs to consider the following factors:

- Functionality needed: will it do the job it needs to do? Is it a suitable material?
- Aesthetics: will the material give the right look and texture to the product?
- Environmental considerations: does the material cause any harm to the environment throughout its lifecycle?
- Availability and cost of stock forms: is the material affordable within the budget set for the product?
- Social, cultural and ethical considerations: have all potential issues surrounding the material's lifecycle been addressed?

Sources and origins

Sources, extraction and conversion

Paper

- Paper is made from wood fibres called **cellulose**.
- Wood is collected, de-barked (the bark is removed) and chipped into small pieces.
- It is then either ground (in the mechanical pulping process) or chemicals are used (in the chemical pulping process) to remove the lignin (the natural 'glue' that holds the fibres together) to create pulp.
- Further chemicals can be added to the pulp to bleach or colour the paper and add texture.
- The pulp is then sprayed in thin layers on to fine mesh, before being compressed and dried out by running it through heated rollers.

> **Cellulose:** Wood fibres, an organic compound structurally important in plant life.

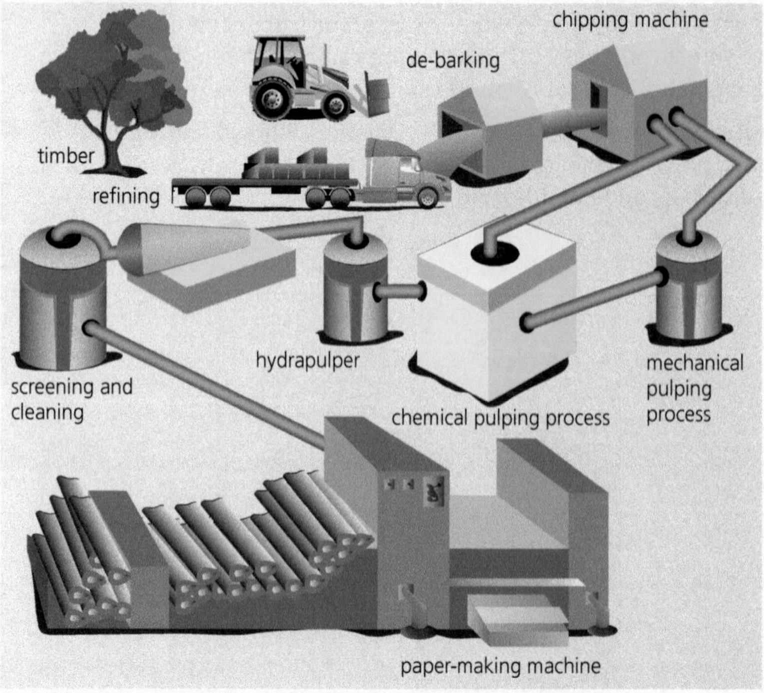

Figure 15.2 **The paper-making process**

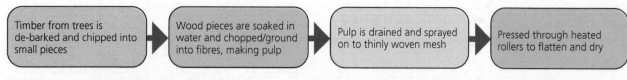

Figure 15.3 The mechanical pulping process

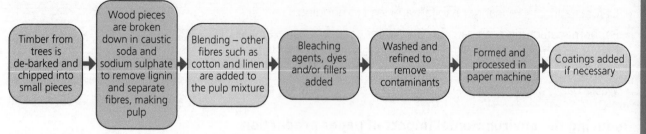

Figure 15.4 The chemical pulping process

Recycled paper

- Recycled paper is made by soaking and mixing waste paper in water to separate the fibres back into pulp. It is then refined to remove contaminants before being sprayed and compressed in the same way as virgin fibre paper.

Card

There are different ways to produce cardboard:

- Multiple layers of paper can be sandwiched and pasted together.
- Layers of wet pulp used for making paper can be pressed together into a thicker layer.

Corrugated card

Corrugated cardboard is made by passing paper through a corrugation machine, which has three layers.

- High-pressure steam is applied to the centre layer soften the fibres before they are pressed to the required thickness and crimped.
- The two outer layers are glued on each side of the wavy centre layer.
- The corrugated cardboard is then cut into large pieces (called 'blanks') that are printed, cut and glued together using other machines.

Foam board and Styrofoam

- Foam board is produced by sandwiching foam between two layers of paper or thin card.
- Expanded polystyrene foam (Styrofoam) is made by heating small pellets of polymer (to approximately 200 °C) and allowing the gas in the polymer to escape and air to enter.

Corriflute

Corriflute is manufactured by drawing molten polypropylene through a former to mould it into a corrugated shape.

Ecological, social and ethical issues

There are many ecological, social and ethical issues associated with paper and card processing:

- A large number of trees are felled to produce wood pulp for virgin paper. Trees are vital for absorbing carbon dioxide in the atmosphere.
- Deforestation can endanger habitats and lead to flooding.
- Paper manufacturing can cause air and water pollution. For example, hydrogen peroxide and sodium hydrosulphite, which are used as bleaching agents, can be harmful to the environment.
- Waste paper accounts for a large percentage of the waste we produce.

Reducing the environmental impact of paper production

- The Forest Stewardship Council (FSC) manages forests used for paper to ensure that one tree (or more) is planted for every one felled. Using paper from FSC forests helps to ensure a continuing supply of trees.
- Lignin extracted in the pulping process can be burnt as a fuel oil substitute in some manufacturing plants, reducing energy demands. Burning bark and other residues can also be used to supply power.
- Using recycled paper means fewer trees are needed to make paper, reducing the demand for fresh pulp and lowering the energy needed to produce the pulp.
- Producing less pulp also reduces the amount of air and water pollution produced during pulp manufacture.

Lifecycle

Paper and cardboard **decompose** more quickly than other paper and board materials.

> **Decompose:** To decay and become rotten.
>
> **Landfill:** The disposal of waste matter by burying it.

Table 15.4 Time taken to decompose for different board materials

Material	Time taken to decompose
Paper	2–4 weeks
Cardboard	2 months
Styrofoam	50 years
PVC	500 years

- Paper and cardboard do not release harmful chemicals when they decompose, unlike laminated layers containing Styrofoam or PVC, which can release toxic chemicals.

Recycling, reuse and disposal

Paper and cardboard

- Most (approximately 70 per cent) of paper and cardboard products are recycled. It is the most commonly recycled material.
- There is a limit to how many times paper can be recycled, as the fibres get shorter and weaker each time. At the end of their useful life they are often burned as a fuel source, or left to decompose.
- Some papers and cards cannot be recycled; these end up in **landfill**.
- Food packaging is often not recycled because it is contaminated by grease and oil.
- An alternative to landfill for non-recyclable paper and cardboard is composting. This has the added benefit of creating mineral-rich soil.

Figure 15.5 Card and paper can be composted in a compost bin

Laminated boards

- Laminated boards such as Styrofoam, PVC foam and Corriflute are much harder to recycle because they are made of polymers that need to be sorted, cleaned, chipped and melted down before they can be used.
- The recycling process can also use a lot of energy.

Foam board usually ends up in landfill because its outer layers of paper must be separated from the foam core before recycling, which is difficult and time-consuming.

Commonly available forms and standard units of measurement

REVISED

Paper and board sizes

- Paper and boards are available in standard-sized sheets ranging from A10 through to A0.
- The most common sizes used by designers are between A6 and A0.

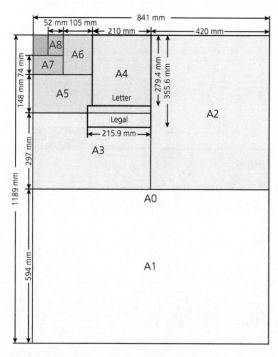

Figure 15.6 **Common paper sizes**

Laminated board sizes

Table 15.5 **Laminated board sizes**

Board	Sizes	Thicknesses	Finishes
Foam board	Standard sheet sizes from A4 to A0	3 mm, 5 mm, 10 mm	
Corriflute	A range of sizes up to 8 ft by 6 ft	2–10 mm	A range of different colours
PVC foam	Standard paper sizes and larger sizes up to 10 ft by 7 ft	1 mm, 2 mm, 3 mm, 4 mm, 5 mm, 6 mm, 8 mm, 10 mm, 13 mm, 15 mm, 19 mm, 25 mm	Standard and special designer colours; gloss or matt finishes
Styrofoam	Sheet form or blocks	5–165 mm (in increments of 5 mm or 10 mm)	

Standard components

REVISED

Components can be used to join materials together, either temporarily or permanently.

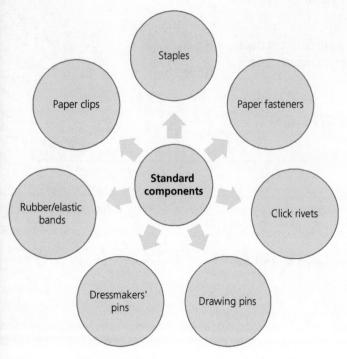

Figure 15.7 Standard components for paper and board products

Structural integrity

REVISED

The structural integrity of a product is its ability to hold together under a load without bending or breaking.

● A folded sheet of paper or thin card is stronger than a flat sheet because the fold creates a rigid section that can support weight.
● Curving or bending paper and card into curved shapes gives it strength and structural stability.
● The wavy centre section in corrugated card provides rigidity and strength.

Structural integrity can be increased by reinforcing one material with another. For example, the layer of foam inside a foam board allows the outer paper layers to stand on their edge and not bend as easily. On their own, neither the sheets of paper nor the foam would have structural integrity or strength. The load is spread across the foam.

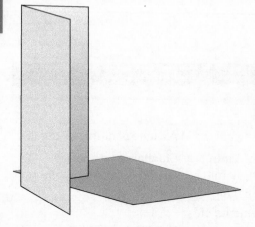

Figure 15.8 A folded sheet of paper is much stronger than a flat sheet

Finishes and surface treatments

The properties and aesthetics of paper and board can be altered and improved by applying different surface treatments.

Paper

Table 15.6 **Characteristics and uses of different paper finishes**

Paper finish	Characteristics	Uses
Cast-coated powder	Provides the highest gloss surface of all coated papers and boards	Labels, covers, cartons and cards
Lightweight coated	A thin coated paper that can be as light as 40 gsm	Magazines, brochures and catalogues
Silk or silk-matt finished paper	Smooth, matt surface; high readability and high image quality	Product booklets and brochures
Calendered or glossy paper	Glazed shiny surface; can be coated or uncoated	Colour printing
Machine-finished paper	Smooth on both sides; no additional coating applied after leaving the paper-making machine	Booklets and brochures
Machine coated	Coating applied while it is still on the paper machine	All types of colour print
Matt-finished paper	Slightly rough surface that prevents light from being reflected; can be coated or uncoated	Art prints and high-quality print work
Supercalendered paper	Gives paper an even smoother surface with a very high lustre	Glossy magazines and high-quality colour printing

Card and board finishes

Varnish

- A thin coating of matt, silk or gloss varnish that can be applied to paper or card products to give them a high shine finish – enhancing their look and feel, and making them last longer.
- Spot UV varnish is a special varnish that is applied, then cured or hardened by ultraviolet light. The varnish makes the coated area shinier and clearer than the surrounding uncoated areas, making parts stand out.

Hot foil application (foil blocking)

- This is used to produce metallic, glossy finishes (for example gold, silver or **holographic** foil).
- It is often used for lettering on invitations and business cards.
- Foil is fed between the card and a magnesium or copper die, which is pressed against the card and releases foil on to the card where the die touches.

> **Holographic:** A special type of photograph or image in which the objects look three dimensional rather than two dimensional.

Embossing and debossing

- Embossing creates a raised area on the paper or card that stands out slightly.
- Debossing creates a sunken or lowered area.
- Two metal formers (a male and a female) in the shape of the design required are used. The card is placed between them and heat and pressure are applied to squeeze and deform the material into shape.

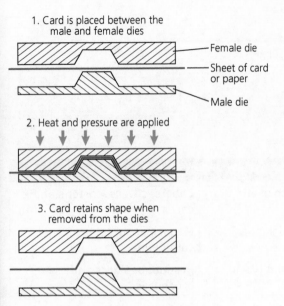

1. Card is placed between the male and female dies

Female die

Sheet of card or paper

Male die

2. Heat and pressure are applied

3. Card retains shape when removed from the dies

Figure 15.9 **The process of embossing**

Laminating

Laminating involves applying a film of clear plastic to one or both sides of paper or thin card.

There are three methods of laminating a document:

- Pouch lamination: pouches are coated on the inside with a thin layer of glue; the document is placed inside and the pouch is heated, activating the glue, which seals the pouch together as it is pressed through rollers.
- Thermal or hot lamination: a commercial lamination method that uses rolls of thin heat-sensitive polymer film.
- Cold lamination: film that has a thin coating of pressure-sensitive adhesive applied to one side is placed over a document and passed through rollers, which press down and smooth out the adhesive, sticking the film to the document. Inexpensive method used for photographs.

> **Typical mistake**
>
> Laminating and varnishing are not the same finish. Laminating is commonly used in schools to put a 'glossy' cover on card or books; it is rarely used in industry. A UV (ultraviolet) varnish would be applied all over, or as a 'spot varnish', to apply a glossy surface finish.

Processes used to make early iterative models

REVISED

- Paper and thin card are used for model making because they can be easily folded, cut, bent, printed on and stuck together.
- Nets are often drawn or printed on to paper or card and then scored, folded, cut out and joined to form the 3D shape required.
- Foam board is used for making architectural models because it is thick and can be stood up and joined easily. It can be cut, scored, indented, printed or drawn on to achieve architectural effects.
- Styrofoam can be shaped and sanded easily and used to form lightweight building blocks that can be joined together to form larger models. It is often used as a mould for vacuum forming or GRP.

Figure 15.10 **Foam board is a good material for architectural models**

Accurate marking out methods

REVISED

- Pencil or pen can be used to mark out on paper, card and foam board.
- A thin permanent marker can be used on Styrofoam, Corriflute and PVC foam.

Stencils and templates

- Stencils and templates can be used for batch production of paper and board products. A shape can be drawn out once on card and cut out to create a cardboard template that is used again and again.
- If multiple numbers of the same part are required, using a template ensures that every piece is exactly the same and saves the time of drawing each part out individually.
- Arranging template pieces so that the material is used to its maximum capacity is known as tessellation. It helps to reduce waste.

Manipulating and joining   `REVISED`

Cutting

- Scissors and craft knives are used for cutting paper and card.
- Foam board, PVC foam and Corriflute can also be cut using a craft knife.
- Thicker Styrofoam (over 10 mm thick) can be cut using a serrated-edged knife, band saw, hacksaw blade or hot wire cutter. It should then be sanded and smoothed with abrasive paper.
- A laser cutter can be used to cut card, PVC foam or Corriflute.
- Avoid using a laser cutter with foam board as fumes are given off and there is a risk of fire.

Folding

- Before folding paper and card by hand, use a blunt knife blade or dull-pointed object to create a clean, sharp crease.
- Hinge cutting is where foam board is cut partway through so that the bottom layer of card acts as a hinge, allowing the card to be folded.
- Vee cutting (where a 'V'-shaped cut is made in foam board and the material removed) allows foam board to be folded inwards and gives a clean, tidy fold.

Joining: adhesives

Different adhesives are suitable for different products and purposes.

Table 15.7 Common adhesives and their properties

Adhesive	Medium	Properties
Glue sticks	Paper and thin card	Safe, easy to apply, inexpensive and mess-free; can unstick over time, pieces can break off, making surface lumpy; quick setting
Spray glues	Paper and thin card	Gives a light, even coating; can be permanent or temporary; quick setting; can be messy
PVA glue	Thicker card, corrugated cardboard, foam board, Styrofoam	Dries clear; inexpensive; can be watered down; only a thin layer is needed for a strong, long-lasting bond
Hot glue guns	Thick card, Corriflute (Must not be used on foam board of Styrofoam as the solvent will melt the foam)	Quick setting; can burn material and user; cool melt versions available; coloured glue sticks available

Adhesive	Medium	Properties
Cyanoacrylate glue (Superglue)	Corriflute, PVC foam (Must not be used on foam board or Styrofoam as the solvent will melt the foam)	Quick setting; very strong bond; short shelf life (approx imately one year)
Polystyrene cement	Corriflute, PVC foam	Dries quickly; clear finish
Contact adhesive	Corriflute, PVC foam	Must be applied to both surfaces; needs to partly set before joining; instant bond; does not require clamping
Epoxy resin	Corriflute, PVC foam	Needs to be mixed with hardener; very strong bond; gives off strong fumes

Digital design tools

REVISED

CAD and CAM

- Computer-aided design (**CAD**) uses a computer to produce designs instead of, or as well as, drawing by hand.
- CAD packages can be used to produce detailed working drawings in different 2D and 3D projections, as well as to simulate how the design will work to a high degree of accuracy.
- CAD drawings can be used to create products using computer-aided manufacture (**CAM**).
- A wide range of CAM machines are available that can create products from paper and boards.
- 2D CAM machines can cut, score and engrave different sheet materials.
- Some CAM machines can mill and form 3D shapes out of Styrofoam.

> **CAD:** Computer-aided design.
>
> **CAM:** Computer-aided manufacture.

Vinyl cutters

- Vinyl cutters are 2D machines that transfer CAD line drawings to self-adhesive vinyl.
- Rollers move the vinyl sheet backwards and forwards on one axis.
- A blade moves horizontally back and forth on the opposite axis, cutting through the top surface of the vinyl and leaving the backing sheet intact.
- The finished design can be peeled off.

Laser cutting

- Laser cutters are usually used for cutting sheet materials, but they can cut a wide range of other materials.
- They cut more quickly and accurately than cutting by hand.
- A laser moves across the sheet material along the x- and y-axes to cut the material.

> **Exam tip**
>
> You need to be familiar with the different materials, their uses and the different manufacturing methods. A laser cutter is not always the most suitable method of production.

Figure 15.11 A laser cutter can cut sheet materials into intricate designs

Milling machines

- Computer numeric controlled (**CNC**) milling machines use a rotating cutter to remove material from the surface of a workpiece.
- Milling machines are mainly used for shaping aluminium and other metals, but can also be used to shape and mill Styrofoam.

> **CNC:** Computer numeric control.

Manufacturing at different scales of production and processes used

REVISED

Paper and board products can be made using different methods depending on the quantities required.

Scales of production

Table 15.8 **Scales of production**

Scale of production	Description	Examples
One-off, bespoke production	A production method used to make a single product. Used to create one-off prototypes. Can be labour intensive and time consuming.	Handmade birthday card
Batch production	A production method in which a limited number of items are produced in one go. Allows for the production of similar items with variations (e.g. colour or text changes).	Packaging where the net (development) is the same shape for each product but the graphics vary, e.g. limited edition flavour cereal boxes
Mass production	A production method used when a very large number (possibly thousands or millions) of a product is produced.	Newspapers and magazines
Lean manufacturing	A manufacturing strategy based on reducing waste at all stages of production.	Can be applied to any product by reducing set up or changeover times during the manufacturing process, reassigning staff to other jobs during any downtime, or reducing the range of products available
Just-in-time (JIT) production	A manufacturing method in which materials are ordered to arrive just in time for manufacturing to begin.	Often used by small businesses, for example those specialising in on-demand merchandise, publishing and printing for products such as business cards, T-shirts, flyers and badges

Manufacturing processes for large-scale (mass) production

Commercial printing methods are used when large numbers of paper and board products are needed.

Offset lithography

- Offset lithography is a commercial printing method based on the principal that oil and water do not mix.
- An image is transferred on to a printing plate that is designed to attract ink and repel water.
- The non-image areas of the plate are treated so that they attract water and repel the ink, so that the ink will only stick to the required areas.
- Offset lithography uses four ink colours, which are overlaid to create others. The colours are cyan, magenta, yellow and black – often referred to as CMYK.

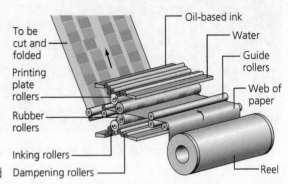

Figure 15.12 Offset lithography printing

Screen printing

- Screen printing is often used for the batch production of repeating patterns or designs, such as on wallpaper or fabrics.
- It uses a stencil placed over a porous fabric mesh screen, which is then stretched over a wooden frame and placed on to paper (or fabric).
- A squeegee is used to spread ink over the screen and force it through the mesh. This process is repeated so that all areas of the material are printed.

Digital printing

- Digital printing is the most cost-effective printing method used in batch production. It allows variations to be made easily and is therefore often used for personalised items in which the design is the same but small details change.
- There are no set-up costs and therefore it is usually used for small print runs.

Die cutting

- Die cutting uses a sharp, shaped blade (the die), which is lowered on to the material. A press is applied to force the blade through the material.
- Die cutting is used to cut many identically shaped products.

Flexography

- Flexography is a mass-production printing process that uses water-based inks.
- Water-based inks dry much faster than oil-based inks, so this is a quicker and cheaper printing process.
- The print quality of flexography is not as good as other printing methods and therefore it is used to print items where quality is less important, such as cardboard boxes, sweet wrappers and plastic carrier bags.

Cost and availability considerations

As costs of materials usually reduce the more you purchase, it is usually more economical to buy a larger sheet of paper or board.

Table 15.9 Costs and availability of materials

Material	Availability	Cost
Paper and cardboard	Widely available	Inexpensive: • A4 white paper – 1p per sheet • Larger sizes (A1 and A0) – 10p per sheet
Cardboard and glossy paper types	Widely available in high street stationery shops	More expensive than standard paper but less expensive than other materials (e.g. metal, timber, polymers)
Foam board	Widely available in sizes A4 to A1 at art and hobby shops	• A4 – 50p per sheet • A1 – £2.50 per sheet Thicker sheets are more expensive
PVC foam	Widely available from specialist suppliers and online	• Small sheets (A4) – £2 per sheet • Large sheets (2,400 × 1,200 mm) – £50 per sheet
Styrofoam	DIY stores and online	Standard 25 mm-thick sheet (2,400 × 600 mm) – £7 per sheet

Example

A designer needs 13 pieces of A4-size foam board to make a prototype product:

- one A4 sheet of foam board costs £0.50
- 13 pieces @ £0.50 = £6.50
- one sheet of A1 size foam board costs £2.50
- eight A4 pieces of foam board can be cut from an A1 sheet

It would be cheaper to buy two A1 sheets and cut each one into eight A4 pieces to make 16 A4 pieces; the designer needs 13 pieces so there would be three pieces spare.

Now test yourself

1 What are 'new wood' fibres commonly known as? [1 mark]
2 What term is used to describe the thickness of card or board? [1 mark]
3 Recycled paper can be recycled indefinitely, true or false? [1 mark]
4 Describe one way that the structural integrity of paper or board can be improved. [1 mark]
5 State the full names of the four colours used in offset lithography and flexography printing. [4 marks]

16 Natural and manufactured timber

Physical and working properties

Timber is a natural, organic, sustainable and renewable material. It is used in many different applications, from building construction, furniture making and boat building to musical instruments and decorative items. Timber is a strong and versatile material with good insulating properties, but different types of timber have different and individual physical and working properties that make them ideal for use in different situations.

Timber is classified into three different types: hardwoods, softwoods and manufactured board.

Hardwoods

Table 16.1 Types of hardwoods

Name	Physical and working properties	Typical uses
Oak	Heavy, hard, tough, open grain, finishes well, good outdoors, contains tannic acid that corrodes screws and stains wood blue	Garden furniture, doors, floors, high-end furniture
Mahogany	Easy to work, available in wide planks, polishes well	Furniture, shop fittings, boat building, doors, pool cues
Beech	Very tough, hard, close and straight grained, hardwearing, easy to work, polishes well, prone to warping	Furniture, toys, wooden tools, good for steam bending
Ash	Wide grained, tough, flexible, finishes well, light in colour	Flooring, tool handles, sports equipment, wooden ladders
Elm	Tough, flexible, durable, water resistant, difficult to work, liable to warping	Boxes, baskets, hockey sticks, archery bows, furniture, wood turning
Teak	Hard; durable; straight grained; easy to work; contains natural oils resistant to moisture, fire, acids and alkalis; expensive	Ship decks, laboratory benches, high-end furniture
Balsa	Soft; lightweight; coarse and open grained; very easy to shape, sand, glue and paint	Model making, packing cases, surfboards, fishing floats

Softwoods

Table 16.2 Types of softwoods

Name	Physical and working properties	Typical uses
Scots pine (redwood)	Straight grain, easy to work, knotty, durable, finishes well, widely available, relatively cheap	Interior construction work, boxes, crates, flooring, paper (pulpwood)
Red cedar	Lightweight; soft; knot free; straight grained; very durable; resistant to rot, weather and insect attack	Outside joinery, building cladding, bathroom and kitchen furniture, wall panels
Spruce	Straight grained, easy to work, fairly strong, low in weight, contains resin canals, some varieties have good resonance	Decorative veneer, flooring, interior construction, musical instruments (piano sound boards, bellies of violins and guitars)
Parana pine	Hard, straight grained, fairly strong and durable, easy to work, tendency to warp	Veneer, furniture, flooring, staircases

Manufactured board

Table 16.3 Types of manufactured board

Name	Physical and working properties	Typical uses
Plywood	Strong and tough due to its layered construction, high strength–to–weight ratio, relatively stable under changes in temperature and moisture, easy to cut but can splinter	Structural panelling in building construction, furniture making
Flexible plywood (Flexi-ply)	Bends extremely easily, easier to cut than normal ply but can still splinter	Curved furniture, shop fittings
Marine plywood	Same properties as plywood but also resistant to moisture	Boat building, decking
MDF	Easy to machine, paint, glue and stain; smooth; chips easily; poor moisture resistance	Furniture, interior panelling, interior doors
Moisture-resistant MDF	Same properties as normal MDF but resistant to moisture, green in colour	Furniture, panelling in kitchens and bathrooms
Flame-retardant MDF	Same properties as MDF but resistant to fire, pink or blue in colour.	Interior panelling and doors where there is above average fire risk
Blockboard	Resistant to warping, relatively easy to cut and finish, requires edging strips when cut	Interior shelving and worktops
Chipboard	Easy to cut, lightweight, breaks and snaps easily, rough finish so usually veneered or covered in plastic, inexpensive	Kitchen worktops, cabinets and shelving (when veneered or plastic coated)

Factors that influence selection

The quality of timber – that is the characteristics that determine its suitability for a particular end-use – can be assessed by grading rules. The effects of specific gravity, knots, wane, slope of grain, rate of growth, **seasoning** defects and blue-stain on quality are discussed below, with special reference to softwoods grown in the British Isles. The influence of pruning is also considered.

- **Suitability** – The physical and working properties of some timbers make them more suitable in different applications, for example teak's moisture-resistant properties make it most suitable for use on ships.

- **Processability and workability** – Some timbers are much easier to cut, plane, shape and sand than others. A timber that is easy to work with will make manufacturing a product much easier and faster.

- **Appearance** – For decorative items such as furniture and musical instruments, the appearance of the product is very important. Timbers such as bird's eye maple and walnut are often selected for their decorative and interesting grain patterns.

- **Durability** – The ability of timbers to be hard wearing and long lasting is important in applications where they will be subjected to constant wear, for example in flooring, or exposed to water or the elements, such as boats and outdoor furniture. Oak is often chosen for its durability.

- **Sustainability** – Designers have a duty to use materials that are sustainable and environmentally friendly. Hardwoods are less sustainable than softwoods as they take much longer to grow. Most softwoods

> **Exam tips**
>
> All woods are easier to work than metals.

> **Seasoning:** Carefully drying woods ready for use.

FSC Wood

come from **managed forests** where new trees are planted for each tree that is cut down.

- **Availability** – Some timbers (particularly hardwoods) are now in short supply and difficult to obtain. Designers and manufacturers may need to choose a different timber with slightly less suitability in order to continue manufacture and meet customer demand.

- **Price (affordability)** – The cost of materials is one of the main factors when designing and making any product. Using costly timber will increase the price of the final product. Manufacturers will often try to buy the cheapest timber they can. Manufactured boards are often used as a cheaper alternative to natural wood.

- **Quality** – The quality of timber can vary hugely because it is a naturally occurring product and no two pieces are exactly the same. The conditions where the timber was grown can drastically affect its quality. Timber is graded to assess its quality and suitability for use. Grading rules assess the effects of:
  - specific gravity
  - knots
  - wane
  - slope of grain
  - seasoning defects
  - pruning.

> **Managed forest:** A forest in which more trees are planted to replace every tree that is felled.

Sources and origins

REVISED

Table 16.4 Sources and origins of softwoods and hardwoods

Softwood	Sources
Redwood (Scots pine)	Northern Europe, Scandinavia, Russia, Scotland
Western red cedar	Canada, USA
Parana pine	South America
Whitewood (spruce)	Northern Europe, Canada, USA
Hardwood	Sources
Beech	Europe (including UK)
Ash	Europe (including UK)
Elm	Europe (including UK)
Oak	Europe (including UK), Russia, Poland
Mahogany	West Africa
Teak	Myanmar (Burma), India, Thailand

Extraction and conversion

Trees are extracted by cutting them down or 'felling'. The process of conversion is as follows:

- The branches and bark are stripped from the trunk.
- They are then sawn roughly into boards or planks using plain sawing and quarter sawing.

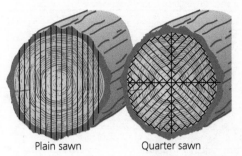

Plain sawn   Quarter sawn

Figure 16.1 Plain and quarter-sawn wood

Seasoning

Once cut into planks, the timber needs to be dried out or 'seasoned'.

Natural seasoning

- Natural seasoning is the cheapest method of seasoning but can take several years.
- The planks are stored outside but protected from sunshine and rain.
- Planks are stacked apart to allow air to circulate between them.

Kiln seasoning

- Kiln seasoning is much quicker than natural seasoning but more expensive.
- Planks are stacked apart inside a large heated kiln.
- The heating and moisture conditions inside the kiln are carefully controlled.
- Kiln seasoning also kills any insect eggs in the wood.

Ecological, social and ethical issues associated with processing

Deforestation

- Advances in technology means machines can now cut down and process trees extremely quickly.
- Acres of forest can be felled and processed in a few hours.
- This has led to a shrinking of the world's forests, known as **deforestation**.
- The consequences of deforestation include:
  - soil erosion, leading to landslides and flooding
  - destruction of animal and wildlife habitats
  - reduction of oxygen and increased carbon dioxide produced

Managed forests

- The Forest Stewardship Council (FSC) promotes and looks after managed forests.
- Managed forests means replanting and maintaining the number of trees and preventing deforestation.
- Timber that is FSC certified is eco-friendly and sustainable.

Exam tips

Managed forests can reverse deforestation.

Deforestation: The over-harvesting of trees creating areas of bare land.

Lifecycle

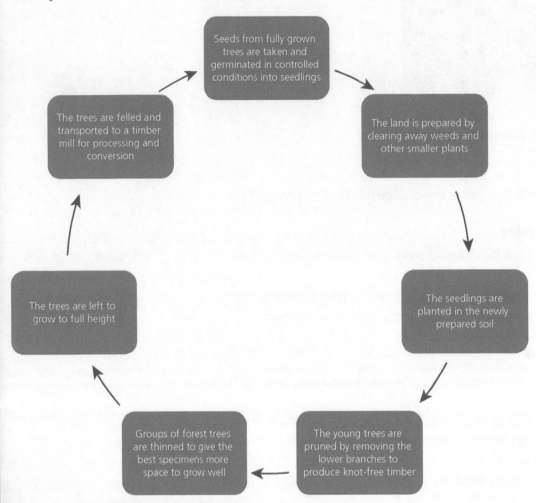

Figure 16.2 **The lifecycle of a tree in a managed forest**

Recycling, reuse and disposal

Reclaimed timber

- Timber that has already been used can often be reused or **reclaimed** to make other products.
- Nails, screws and any other fixings are removed and then the timber is cut up and used for other products.
- Reclaimed timber works best with large items such as railway sleepers, roofing beams or joists.
- Using reclaimed timber to make new products is called 'upcycling'.
- Reclaimed timber is becoming increasingly popular and trendy in designer furniture making.

> **Reclaimed timber:** Old timber that is reused in a new way.

Manufactured boards

- Manufactured boards are made from recycled wood and reduce the amount of natural timber being used.
- Wood veneers, chips, dust and fibres are combined using different methods to make different types of boards.
- Manufactured boards are more stable than natural timber and less likely to warp or twist.
- Manufactured boards are generally less expensive than natural woods and are available in large sheets.

Now test yourself answers and quick quizzes at **www.hoddereducation.co.uk/myrevisionnotes**

Other uses

- Wood that is not of a suitable size or quality to be reclaimed or used for manufactured boards can be cut into chips and used in other products such as:
  - animal bedding
  - mulches, compost and coverings for gardens and landscaping
  - forest paths, arenas, and so on.

Wood as a fuel

- In recent years the use of wood as a fuel has begun to increase.
- A growing number of schools, public and commercial buildings and homes are installing wood burners or wood pellet heating systems.
- Many coal-fired power stations have had some of their boilers converted to burn wood.
- The rising cost of fossil fuels and increasingly tougher environmental laws mean that the use of wood as a fuel will continue to increase.

> **Typical mistake**
>
> Remember the following:
> - Manufactured boards are recycled products but are difficult to recycle because of the adhesive used.
> - Reclaiming wood is a secondary recycling method, whereas manufactured boards are tertiary recycling.

Commonly available forms and standard units of measurement

REVISED

Commercial forms and sizes of timber

- Once it has been seasoned, timber is then cut and planed down into smaller sizes and lengths.
- Timber is sold either rough sawn, planed both sides (PBS) or planed all round (PAR).
- Planed timber is around 3 mm smaller than rough sawn because planning removes around 1.5 mm from each side.
- Timber is usually available in lengths up to 3.0 m from DIY stores but some timber yards can supply lengths up to 4.8 m

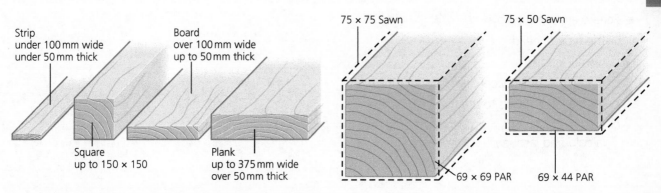

Figure 16.3 Diagram of rough sawn and planed timber

Standard shapes

- Planks – usually around 50 mm thick and 225–375 mm wide.
- Boards – less than 40 mm thick and 100 mm wide or more.
- Strips – less than 50 mm thick and 100 mm wide.
- Squares – smaller than boards but with the same height and width.
- Dowelling – cylindrical lengths of timber usually 40 mm or less in diameter.
- Mouldings – decorative sections of wood made from strips, so usually less than 50 mm thick and 100 mm wide.

Commercial forms and sizes of manufactured boards

- Manufactured boards are available in large sheets measuring 2440 × 1220 mm.
- Many DIY stores supply smaller sizes and can cut sheets in store to customers' requirements.
- Boards are available in 3 mm increments from 3 to 25 mm thick (3, 6, 9, 12, 15, 18, 21, 25).
- Mouldings made from MDF are also available in many different shapes and sizes.

Standard sections

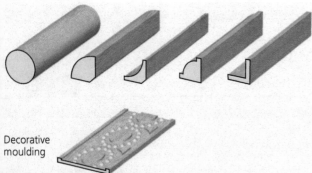

Decorative moulding

Figure 16.4 Different moulding sections

Standard components

Standard components are available for use when constructing and assembling products made from timber and manufactured board.

Woodscrews

- Screws are specified by:
  - the type of head
  - overall length
  - material – usually steel or brass
  - gauge – this relates to the screw's diameter
  - finish – galvanised, zinc coated, chrome plated or japanned (a coloured lacquer).

Table 16.5 **Common types of woodscrew**

Countersunk head	Round head	Raised head	Coach	Twinfast
Used when the head of the screw needs to be flush (level) with the surface of the timber.	Used to fasten brackets and thin sheet materials such as metal and plastic to wood.	Used for decorative purposes such as fitting door handles to timber.	Used when an extremely strong fastening is required. Square head has no slot and is tightened with a spanner.	Used specifically in chipboard. Has two threads for greater holding ability.
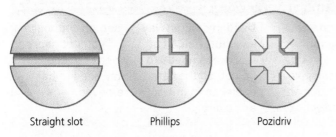 Countersunk	Round	Raised	Coach	Twinfast

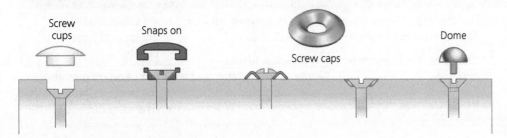

Straight slot Phillips Pozidriv

Figure 16.5 **Different screwdriver slots**

Various caps and covers can be used with screws to enhance the look or 'hide' them from view.

Screw cups Snaps on Screw caps Dome

Figure 16.6 **Caps and covers**

Nails

Table 16.6 **Different nail types**

Round wire nails	Oval wire nails	Panel pins	Masonry nails
Used for general joinery work. 12–150 mm in length.	Used for interior joinery work. Small head makes them easy to hide. 12–50 mm in length.	Used for pinning sheet materials and small-scale work. 12–50 mm in length.	Used when fastening into brickwork or mortar (usually in building construction).

Other fastenings

Table 16.7 Other fastening types

Clout nails	Staples	Cut tacks	Hardboard pins	Corrugated fasteners
Used to fasten roofing felt to shed roofs. Galvanised so will not rust.	Used to make packing crates and in upholstery. Can be hammered in or fired in with a staple gun.	Used to fasten fabric to wooden frames in upholstery.	Used to fasten hardboard to wooden frames, such as wardrobe and cupboard backboards. Diamond–shaped head makes them hidden once hammered in.	Used to make quick corner joints in wooden frames by hammering across the joints.

Hinges

Table 16.8 Types of hinges

Butt hinges	Piano hinges	Butterfly hinges	Flush hinges	Barrel hinges
Used for doors, windows, boxes etc. Need to be recessed into the wood.	Used on piano lids and other situations where lots of support is needed. Can be cut to required length.	Used on lightweight doors and lids on decorative items such as jewellery boxes.	Used in similar way as a butt hinge but no recess needs to be cut, so they are easier to fit but not as strong.	Used on cupboard and cabinet doors where the doors may need to be removed.

Catches

Table 16.9 Types of catches

Magnetic catches	Spring catches	Ball catches	Toggle catches
Used on kitchen cabinets and cupboards. Available in different materials such as plastic, brass and steel.	Used in more heavy-duty cupboards and cabinets. Stronger than magnetic catches. Often incorporate two rollers.	Used in household and office doors. Recessed fitting makes them very neat and unobtrusive.	Used to fasten lids of cases and boxes securely. Often used on portable items such as musical instrument cases.

Structural integrity

REVISED

- The structure of timber is made up of hollow, tube-like cells of different sizes and wall thickness.
- The cells are composed mainly of an organic compound called cellulose.
- Timber has high tensile strength (resistance to stretching and breaking under tension).
- Timber has lower compressive strength (resistance to squashing and breaking under compression).
- Hardwoods contain fibrous material and are stronger than softwoods.
- The structural integrity of timber structures can be improved by the use of triangulation.
- Triangulation is the forming of triangular shapes to give stability and strength to structures.

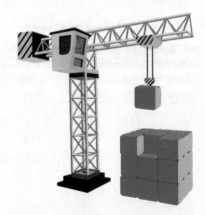

Figure 16.7 A crane jib showing the use of triangulation

Now test yourself answers and quick quizzes at **www.hoddereducation.co.uk/myrevisionnotes**

Knock down fittings

- **Knock down (KD) fittings** allow timber pieces to be joined together quickly and easily.
- Products that use KD fittings can also be quickly and easily dismantled and reassembled.
- They are used for most flat-pack furniture, such as MDF and chipboard wardrobes and kitchen units.
- They can be assembled without glue using a few simple tools such as a screwdriver, allen key and hammer.
- KD fittings improve the structural integrity of the products they are used on.

> **Knock down fittings:** Fittings used in flat-pack furniture to make assembly easy.

Finishes and surface treatments

REVISED

- Finishes have two main purposes:
  - to enhance the appearance of the timber
  - to protect the timber from damage.
- Before applying a finish, the timber surface should be properly prepared by planing or sanding.
- The three types of abrasive for use on timber are:
  - glass paper – ground glass glued on to 280 × 230 mm backing paper
  - garnet paper – hard, crushed stone glued on to backing paper (more expensive than glass paper)
  - wire wool – fine strands of wire used as a light abrasive in the final stages of sanding.
- Abrasives are graded according to the size of the particles on or in them.

Table 16.10 **Grades of abrasive**

Grade	Uses
Extra coarse	Removal of old and/or very stubborn varnishes or paint layers
Coarse	Rough shaping of timber after cutting Removal of paint or varnish
Medium	General sanding work Some final shaping of timber
Fine	Final sanding of timber before finishing
Extra fine	Sanding between coats of paint or varnish to provide a 'key'
Super fine	Very fine sanding between delicate coatings For polishing hardwood

Table 16.11 Knock down fittings

Corner block		• Made from moulded plastic. • Used for joining timber pieces at right angles. • Screws fasten the fittings to the timber pieces. • Used in frames, shelving units, bookcases and small tables.
Two block fittings	Pins Bolt	
Rigid joint		
Connector bolt		• Made from steel or brass. • Bolts and threaded pieces fit into holes in the timber. • As the bolt is tightened the two pieces are pulled together. • Used in beds and bedroom furniture such as wardrobes and drawers.
Cross dowels	Rail Aluminium barrel Allen screw Locating pin Leg	
Cam locks	Screwdriver slot Inside of drawer Threaded rod Drawer front	• Made from steel. • Disc and threaded bolts fit into holes in the timber. • Bolt shaft engages with slot in disc. • Disc is rotated by a screwdriver and locks the shaft in position. • Used in beds and bedroom furniture such as wardrobes and drawers.
Bench top joiners		• Made from steel. • Curved or angled plates are inserted into holes in each piece. • A nut and bolt goes through both plates. • As the bolt is tightened the two pieces are pulled together. • Bench top joiners are used for joining kitchen worktops and benches together
Table plate		• Table plates are used to join large table legs to table tops.

Varnish and lacquer

- Can be oil, water or solvent based.
- Normally transparent and available in matt, satin or gloss finishes.
- Can be sprayed or painted on.

Oil

- Made from 'natural' oils produced by trees.
- Many different types are available: teak oil, Danish oil, Tung oil, linseed oil.
- Easy to apply with a cloth or brush and can be recoated at any time.
- Vegetable oil can also be used for kitchen utensils such as spatulas and chopping boards.

French polish

- Solvent based – made by mixing shellac (a resin created by insects) in methylated spirits.
- Applied by brush or cloth in thin layers that build to give a deep shine.
- Wax is usually applied over the top to further improve the shine.

Wood stains

- Can be water or solvent based.
- Available in a wide range of different colours but must be darker than the wood they are used on.
- Improves/alters the appearance of the timber but does not provide any protection; a varnish or wax coating is usually applied over the stain.

Sanding sealer

- Quick drying, solvent-based liquid.
- Seals the wood surface and raises the fibres of the wood for sanding.
- Applied before a wax or varnish.

> **Typical mistake**
> Sanding sealer is applied to wood prior to sanding not after.

Paint

- Paints can be water, oil or solvent based.
- Can be applied by brush or sprayed on.
- Many different types of paint are available:
  - enamel paint – permanent and durable for use on hard wooden surfaces
  - acrylic paint – easy to apply by brush and often only requires one coat
  - spray paint – quick-drying, solvent-based paint that can cover large areas or hard-to-reach areas easily. Often multiple thin coats are needed as it runs.

Processes used to make early iterative models

Timber and manufactured boards are available in thin sheets (3 mm or less). This makes them useful for creating iterative models and developing ideas.

Table 16.12 Types of timber and manufactured boards, fixing and features

Material	Considerations when model making
MDF	• Can be cut by hand with a coping saw, or by using a fret saw or band saw. • Thin MDF can also be cut and engraved on a laser cutter. • Thin MDF is easy to bend and form into different shapes.
Plywood	• Can be cut with similar tools to MDF. • Splinters easily, especially when bent. • 'Laser ply' is available, which can be used with a laser cutter.
Balsa wood	• Extremely lightweight and easy to shape. • Available in blocks, strips and sheets that can be easily cut and shaped with a craft knife. • Much easier to work with than MDF and plywood but considerably more expensive.
Pre-cut and pre-formed timber	• Small lengths of square and rectangular section softwood and dowel are available in a range of small sizes suitable for modelling. • They can be cut quickly and easily using a junior hacksaw. • Other pre-formed items (such as lollipop sticks, matchsticks and small pre-cut wheels) are available to use for iterative modelling.
Fixing	• Hot glue guns are the quickest and easiest way to join timber when modelling as the glue dries in seconds. • 'Cool' glue guns work in the same way and are safer for younger children. • PVA wood glue can be used for modelling but it can take several minutes or even hours to dry. • Specialist solvent-based glues such as UHU or 'No nails' must be used in well-ventilated areas. Drying times vary depending on the type and amount used. • Other non-solvent based adhesives are available, but these have limited strength and longer drying times.

Manipulating and joining

Handsaws

- 'Handsaw' is the general name given to saws used to cut large pieces of timber by hand.
- Cross-cut saws are used to cut across the grain.
- Ripsaws are used to cut along the grain.
- Tenon saws are used for making straight cuts in timber blocks and strips.
- Coping saws are used to cut curves in thin pieces of timber.

Electric saws

- Electric saws use electric motors to move the saw blade, making cutting timber faster and easier.
- Jigsaws are used for cutting shapes in manufactured board. Different blades are available for fine, medium and coarse cutting. Cutting depth is limited to 30–40 mm.
- Fret saws have thin blades and are used for cutting fine curves in thin timber sheets.

- Band saws are large free-standing saws with a steel blade shaped like a large rubber band. They can be used for cutting straight lines and curves, and can cut very thick pieces of timber.
- Circular saws have a high-speed rotating circular blade and are used for cutting straight lines in long lengths of timber and board.

Drills

Drills can be hand powered, battery powered or electric. They can be hand-held, portable or fixed to the floor.

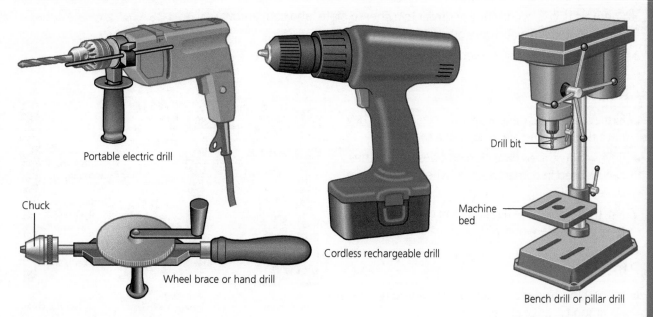

Portable electric drill

Chuck

Wheel brace or hand drill

Cordless rechargeable drill

Drill bit

Machine bed

Bench drill or pillar drill

Figure 16.8 Different types of drills

Drill bits

- Twist bits are the most common type of drill bit and are used for drilling holes in metal and plastic.
- Countersink bits create a tapered hole for screws so that they lay flush (level) with the surface of the timber.
- Forstner bits have large heads for creating large, deep, smooth-sided holes.
- Flat bits are flat and can make fast, accurate holes.
- Hole saws are cylindrical-shaped saws with teeth like a normal saw that rotate. Hole saws remove the centre of the hole in one piece, unlike forstner or flat bits, which destroy the centre of the hole.
- Expansive bits are adjustable and are used to drill shallow holes in wood between 12 and 150 mm in diameter.

Files

- Rough files (called bastard files) are used for coarse work and remove lots of material.
- Second cut files are used for general shaping work.
- Smooth and dead smooth files are used for very fine shaping before polishing.
- Needle files are very small files used for intricate work.

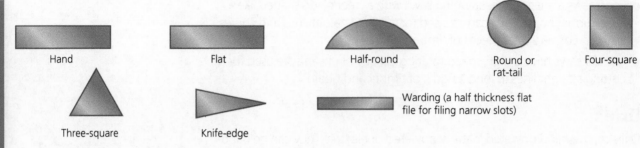

Figure 16.9 **Different file shapes**

- Surforms have a rough surface similar to a cheese grater and can remove large amounts of wood very quickly.
- The blades can be flat, curved or round for forming different shapes.

Chisels

- Firmer chisels are general all-purpose chisels. They have flat sides and are used with a mallet.
- Bevel-edge chisels have sloping sides so that they can be used in corners, such as when cutting dovetail joints.
- Mortise chisels have flat sides and thicker blades to lever out waste wood without breaking.

Planes

- Smoothing planes and jack planes are the two most common types of plane.
- Smoothing planes are around 250 mm long with extremely sharp blades for fine finishing and planing end grain.
- Jack planes are longer and heavier for quick removal of waste wood along the grain of the wood.
- Electric planers are hand held and work in a similar way to jack and smoothing planes but they have a rotating blade, making the removal of wood quicker and easier.
- Planer thicknessers are fixed machines with rotating blades that can plane wood extremely evenly to a specific thickness.

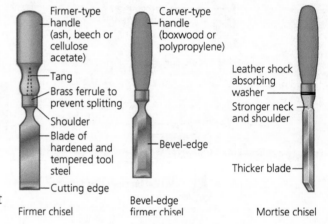

Figure 16.10 **Different chisel types**

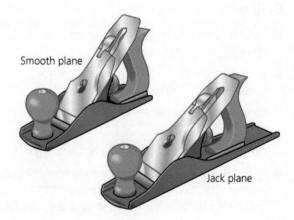

Figure 16.11 **Different types of plane**

Sanders

- Orbital sanders are hand held; abrasive paper sheets are attached with clips and they vibrate in small circular motions.
- Palm sanders are small sanders that can be used with just one hand. The abrasive paper sheets are attached using Velcro and can be round or triangular in shape.
- Belt sanders have two rotating drums that turn a continuous band of abrasive paper at high speed for rapid sanding and shaping of timber. They can be hand held or fixed to the floor and require some form of extraction due to the amount of dust created.
- Disc sanders have a circle of abrasive paper fixed to a spinning disc and can quickly shape timber. Due to the rotation of the disc, only one side of the disc can be used (the other is fitted with a guard).

Lathes

- Wood-turning lathes grip and rotate pieces of timber at high speed.
- Special tools are then used to cut and shape the wood as it turns.
- Wood-turning lathes are used for:
  - Between centres turning – the timber is held at both ends between the headstock and tailstock of the lathe and used to make cylindrical items such as table legs.
  - Faceplate turning – the timber is screwed to a faceplate on the outer end of the lathe and used to make circular items such as bowls.

Routers

- Routers are used for cutting grooves and profiles into timber and manufactured board.
- The shape of the cutter determines the shape of the profile or groove.
- Hand-held routers have guides fitted to help cut parallel to the edges of the timber.
- Table routers are mounted below the work surface, with the cutter protruding upwards.

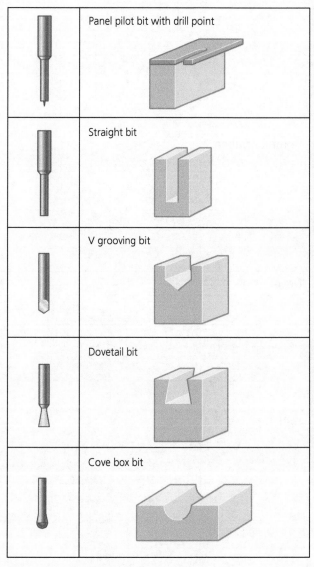

	Panel pilot bit with drill point
	Straight bit
	V grooving bit
	Dovetail bit
	Cove box bit

Figure 16.12 Different router bits

Wood joints

Table 16.13 Types of wood joints

Joint	Illustration	Description	Example uses
Butt		A simple joint in which two pieces of timber are joined by placing their ends together and nailing or gluing them.	Basic boxes or cabinets, building frames
Dowel		A series of corresponding holes are drilled in the joint surface of two pieces of timber. Short dowels are then inserted with glue and the joint is clamped together until dry.	Chair and table legs, table tops, cabinets and panels
Comb or finger		Made by cutting rectangular cuts in two pieces of timber, which interlock at a 90-degree angle when glued.	Tables and chairs, floorboards, roof and door construction
Dovetail		Pins are cut in one piece of timber that interlock with a series of tails cut in another piece of timber. These are glued together.	Joining the sides of a drawer to the front, jewellery boxes, cabinets
Half-lap		Two pieces of timber joined together by overlapping them. Material is removed from each piece of timber at the point of intersection.	Framing
Mitre		A joint made by cutting the two pieces of timber to be joined at an angle of 45 degrees so that a 90-degree corner is formed.	Picture frames, pipes, moulding

Housing	Housing joints Stopped housing	A joint made by cutting a channel in a piece of timber and inserting and securing another piece of timber into the channel.	Fitting shelves and partitions into bookcases and cabinets
Mortise and tenon		A joint comprising of a mortise hole and a tenon tongue. The tenon is cut in one piece of timber (known as the rail) to fit exactly into the mortise hole in the other piece of timber.	Table and chair legs
Bridle		Similar to a mortise and tenon joint but with a longer tenon (as long as the depth of the timber) and a mortise that is cut to the whole depth of the timber so that the two pieces of timber lock together tightly.	Legs or stiles to rails, frames
Corner halving		A joint in which channels that interlock are cut into the corner of two pieces of timber. They are often glued and nailed or screwed to reinforce them.	Frames
Cross-halving		Constructed in the same way as the corner halving joint, but the joint is made in the middle of the timber to allow any internal parts of the frame to 'cross' each other inside the frame.	Strengthening rails for tables and chairs, trellis, box compartment dividers
Biscuit		Small slots are cut into the edges of the boards using a biscuit joiner. An oval-shaped compressed wood fibre biscuit is then glued into the slot and the two boards are clamped together.	Table tops

Bending timber

- Kerfing is the technique of making evenly spaced cuts into a piece of timber, allowing it curve more easily.
- Flexy MDF is available, which is sheet MDF with kerfing on one side, to make it flexible.
- Steam bending involves softening timber by placing it in a steam-filled chamber so that it absorbs the hot moisture and becomes easier to bend.
- The steamed timber is bent and clamped into a former then left to dry thoroughly so it keeps the shape.
- Laminating uses thin layers of wood called veneers that are glued together and clamped around a former until the glue sets.
- Vacuum pressing is similar to laminating but uses a vacuum bag instead of clamps, which is sucked around the former to hold the veneer laminates firmly in place until the glue sets.

Accurate marking out methods

REVISED

Marking out accurately is important to ensure different parts are cut and shaped to the correct size.

Marking out tools

- Marking knife – Used instead of a pencil to mark much thinner and more accurate lines on to timber.
- Try square – Used to mark a line at 90 degrees to an edge or to check the angle of a corner.
- Mitre square – Used to mark a line at 45 or 135 degrees to an edge.
- Marking gauge – Used to draw a line along the grain parallel to the edge of the timber.
- Cutting gauge – Used to draw a line across the grain parallel to the end of a timber piece.
- Mortise gauge – Used to mark out the position of the slot for a mortise and tenon joint.
- Sliding bevel – Used to draw any angles by setting with a protractor and tightening in position.

Digital design tools

REVISED

- Computer-aided design (CAD) is used along with other processes to create various effects and finishes on timber:
  - Laminate flooring made from MDF can be printed with a plastic wood-effect layer on top.
  - Laser cutters can be used to cut a series of thin cuts in pieces of wood to make them bend like a hinge.
- Computer numeric control (CNC) machines are computer-controlled routers that can cut out and carve accurate and extremely detailed shapes from timber in a fraction of the time taken by hand.
- 3D printers that use fine wood particles instead of plastic can make complex shapes that can be sanded and painted in a similar way to wood or manufactured board.

Manufacturing at different scales of production and processes used

Lean manufacturing

Lean manufacturing involves cataloguing waste and off-cut pieces of wood which are then barcoded and stored for later use instead of cutting fresh pieces or buying more wood unnecessarily.

Jigs and fixtures

- Fixtures are devices that hold the workpiece in a specific position so that it can be machined easily and quickly without repeatedly marking out.
- Jigs are devices fixed to pieces of work that guide the tool on to the workpiece in a specific position.
- Jigs and fixtures are usually used together.

Templates and patterns

- Templates are used to mark out complex or irregular shapes repeatedly.
- Paper and card templates can be used but acrylic, MDF or metal templates are often used in manufacturing as they are more hardwearing and longer lasting.
- Patterns are fixed directly to the material as a guide for cutting.
- Paper patterns are often printed and glued on to timber.

Steam bending and laminating machines

- Steam bending machines, which use hydraulic presses to press steamed timber around the former, are used when producing steam–bent products in large quantities
- Laminating presses press and hold laminated shapes together while adhesive dries.
- Roller coaters are machines that apply a thin, even coating of adhesive to timber pieces for assembly, including laminating.

CNC machines

- Multi-axis CNC routers have cutting heads that can rotate to any angle and cut out complex 3D shapes in timber.
- CNC wood lathes have a computer-controlled cutting head that can automatically select different cutting tools and cut identical designs quickly and repeatedly.
- CNC drilling machines are computer-controlled machines that can drill a series of holes to a pre-set depth in timber pieces. They are often used in the manufacture of flat-pack furniture using knock down fittings.

Cost and commercial viability, different stakeholder needs and marketability

When choosing timber for a product, designers and stakeholders need to consider not only the structural and physical properties of timber, but also the cost and availability.

Cost

- Timber prices fluctuate (go up and down) depending on market supply and demand.
- Timber prices also vary depending on your location – timber grown locally is much cheaper as there are fewer transport costs.
- The quality of the timber, such as straightness of grain and how well seasoned it is, will also affect the cost.
- Harvesting costs can vary because:
  - larger trees are less expensive to harvest
  - hardwood trees take longer to season
  - weather conditions can reduce the number of trees harvested.

Availability

- Certain types of timber are scarce and harder to come by or difficult to import, which will affect the production speed and volume of the finished product.
- Some timbers that were once popular but are now not fashionable are not exported as they are unlikely to sell.
- Certain timbers are only available in certain sizes depending on the size of the actual tree.

Other considerations include:

- Environmental factors, such as whether the timber is from a sustainable source.
- Safety concerns – certain woods produce dust during manufacture that can be extremely harmful to humans.

Calculating the quantities, cost and sizes

- To calculate quantities, a cutting list should be drawn up or written out that breaks down the product into the amounts of each size needed.
- Timber is usually sold by the metre or in pre-cut lengths from 1800 to 3600 mm.
- When calculating amounts needed, you should decide which will be cheapest and give the minimum of waste material.

Example:

The shelving unit shown in Figure 16.14 is made from 100 × 18 mm pine.

Table 16.14 Cutting list

Part	Number required	Length (mm)	Total length (mm)
Left side	2	864 mm	1728 mm
Right side	2	864 mm	1728 mm
Top	2	400 mm	800 mm
Base	2	400 mm	800 mm
Shelf	1	364 mm	364 mm
			TOTAL = 5420 mm

100 × 18 mm pine costs £4.59 per metre or £12.99 for a 3.6-metre length.

Option 1: Buy six metres @ £4.59 per metre = £27.54 (you will have 560 mm left over).

Option 2: Buy two 3.6-metre lengths @ £12.99 = £25.98 (you will have 1.78 m left over).

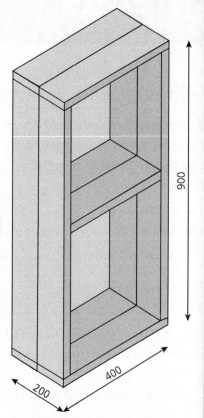

Figure 16.13 Shelving unit

1 Name three ways screws are specified. [3 marks]
2 Describe two benefits of piano hinges. [2 marks]
3 Name three different knock down fittings. [3 marks]
4 Other than painting, name three other finishes suitable for wood. [3 marks]
5 Describe one use of balsa wood. [1 marks]
6 Explain what is meant by 'kerfing'. [1 marks]

17 Ferrous and non-ferrous metals

Physical and working properties

The general characteristics of metals are that they:

- have a high melting point
- have high tensile strength
- have a lustrous shiny finish
- are **malleable**
- are **ductile**
- are good conductors.

Ferrous metals

- Ferrous metals are metals that contain iron.
- All ferrous metals are technically alloys because they are mostly made of iron but contain small amounts of other metals or substances.
- Ferrous metals corrode easily due to their high iron content.
- Ferrous metals are attracted to magnets.
- The exception to this is stainless steel, which is very resistant to rust but still attracted to magnets.

> **Malleability:** A material's ability to deform without breaking or snapping when hammered or rolled into a thin sheet.
>
> **Ductility:** How easily a material can be deformed or bent out of shape without snapping or breaking.

Table 17.1 **The properties, working characteristics and uses of ferrous metals**

Name	Properties/working characteristics	Uses
Low carbon steel or mild steel	• Inexpensive compared to other metals • Tough • East to cut, drill and weld	General building and engineering (nuts, bolts, girders), car body panels, gates
Medium carbon steel	• Similar properties to low carbon steel • Slightly harder and less ductile	Large vehicle parts such as axles, gears and crankshafts
High carbon steel (or 'carbon steel')	• Much harder and stronger than mild steel • More brittle	High-strength cables, springs, saw blades, drills, etc.
Tool steel (die steel)	• Extremely hard • Resistant to heat	Cutting tools, machine parts
Stainless steel	• Resistant to wear and corrosion	Cutlery, surgical instruments, kitchen utensils, specialist vehicle parts
Cast iron	• High compressive strength • Extremely brittle • Extremely resistant to corrosion and oxidisation	Car engine blocks, manhole covers, kitchen saucepans
Wrought iron	• Strong • High resistance to corrosion • Attractive 'patina' as it ages	Ornamental gates, fences, garden benches

Non-ferrous metals

- Non-ferrous metals do not contain iron so are not attracted to magnets and do not rust easily.
- There are over 50 different types of non-ferrous metals.

Table 17.2 The properties, working characteristics and uses of non-ferrous metals

Name	Properties/working characteristics	Uses
Aluminium	- Lightweight - Attractive, natural finish - Malleable - Good conductor of heat and electricity - Can be polished to a mirror finish	Drinks cans, foil, automotive parts, cooking utensils, window frames, ladders, roof joists, body shells of cars and aircraft
Copper	- Soft and extremely ductile - Malleable - Excellent conductor of heat and electricity - Good resistance to corrosion	Electrical cables, water pipes, cooking pots, jewellery, statues, roof coverings
Lead	- Soft and malleable - Dense (heavy) - Good resistance to corrosion - Extremely resistant to acid	Car batteries, fishing and diving weights, weather protective flashing around house roofs
Zinc	- Weak (low compressive and tensile strength) - Brittle - Poor conductor of electricity and heat - Extremely resistant to corrosion - Retains its shiny, silvery finish	As a coating on products made from ferrous metals such as steel (buckets, watering cans, automotive parts, bolts and screws, etc).
Tin	- Ductile - Malleable - Easy to break - Very resistant to corrosion - Retains its shiny appearance when exposed to moisture	As a coating on the surface of other metals to prevent them corroding, for example food cans or 'tin' cans

Alloys

- Alloys can be ferrous or non-ferrous.
- Alloys are metals mixed with other substances to make them stronger, harder, lighter or better in some way.
- Steel (all types) is technically an alloy because it is made from iron mixed with small quantities of other materials.
- Shape memory alloys (SMAs) are special alloys that return to their original shape when heated.
- The most common SMAs (known as nitinol alloys) are a mixture of nickel and titanium and are used in many medical and military products.

Typical mistake

Unlike gold and silver, bronze is an alloy and not a pure metal.

Table 17.3 The content, properties, working characteristics and uses of alloys

Name	Approximate contents	Properties/working characteristics	Uses
Brass	• 65% copper • 35% zinc	• Durable • Good corrosion-resistance	Musical instruments, door knockers, letterboxes, handles, boat fittings
Pewter	• 85–99% tin • 1–15% copper, lead and antimony (a hard, brittle metal with a bright silvery finish)	• Tough • Malleable • Polishes to bright finish	Drinking tankards, jewellery, picture frames, ornaments Often used as a cheaper alternative to silver
Duralumin	• 94% aluminium • 4% copper • 1% magnesium • Very small quantities (less than 1%) of manganese and silicon	• Lightweight but much stronger than aluminium • Extremely strong • Corrosion resistant • Difficult and expensive to produce	Aircraft frames, car chassis, speedboats, door handles, hand rails, etc.
Solder	• 70% lead • 30% tin • Note: Modern solder does not contain lead and is approximately 99% tin and 1% copper	• Malleable • Stronger than lead and tin • Low melting point • Good electrical conductivity	Electrical connections on printed circuit boards and pipework joints in plumbing
Alnico	• 50% iron • 15–26% nickel • 8–12% aluminium • 5–24% cobalt	• Extremely magnetic • Good electrical conductivity • Good corrosion resistance	Magnets in loudspeakers, guitar pickups
Bronze	• 88% copper • 12% tin • Small amounts of other metals such as aluminium, zinc, lead and silicon	• Malleable • Ductile • Good heat and electrical conductivity • Attractive golden-brown colour	Sculptures, statues, musical instruments, trophies, medals
Gunmetal	• 80–90% copper • 3–10% zinc • 2–11% tin	• Tough • Durable • Very low friction (slippery surface)	Valves, hydraulic equipment, bearings, bushes, gear wheels and fittings
Sterling silver	• 92.5% silver • 7.5% copper	• Shiny and attractive • Ductile • Extremely soft • Tarnishes over time if not cleaned	High-quality cutlery, medical instruments, jewellery, musical instruments

Factors that influence selection

- Some metals have particular qualities that make them more suitable than others for particular uses.
- The characteristics of different metals are the main factors that will influence their selection and use.
  - **Suitability** – The physical and working properties of some metals make them more suitable in different applications, for example aluminium's light weight makes it most suitable for use on aircraft.
  - **Processability and workability** – Some metals are much easier to bend, shape and work than others. A metal that is easy to work with will make manufacturing a product much easier and quicker.
  - **Appearance** – For decorative items, such as furniture and musical instruments, the appearance of the product is very important. Metals such as bronze and silver are often selected for their shiny and attractive colour.
  - **Durability** – The ability of metals to be hard wearing and long lasting is of importance in applications where they are likely to be subjected to constant wear, for example tools, blades or building construction.
  - **Conductivity** – Products are often needed that require heat or electricity to be transferred effectively. Although all metals conduct heat and electricity, some are much better than others. Copper is an excellent conductor so is often used for electrical cables and cooking pots.
  - **Availability** – Some metals and alloys are in short supply and much more difficult to obtain than others. Designers and manufacturers may need to choose a different metal, such as an alloy with slightly reduced suitability, in order to continue manufacture and meet customer demand.
  - **Price (affordability)** – The cost of materials is one of the main factors when designing and making any product. Using excessively costly metals will increase the price of the final product. Some alloys are made as cheaper alternatives to other metals, for example pewter – an alloy made from tin and copper – is very similar to silver in appearance but much cheaper.

Sources and origins

- All metals are sourced from metal ores, which are naturally occurring rocks in the Earth's crust.
- The ores contain metal oxides, which are metals that have reacted with air or water.
- Ores are dug out from the ground by mining.

Extraction

- Metal ores must be separated (extracted) from oxides before they can be used.
- Some metals, such as gold, do not always need extraction – these are known as 'native metals'.
- Some metals are harder to extract than others, depending on their reactivity.
- There are three main methods used to extract metals:
  - extraction by chemical reaction
  - extraction by carbon or carbon monoxide reduction
  - extraction by electrolysis.

Table 17.4 Methods of extracting metals

Ease of extraction	Metal	Extraction method used
Hard to extract	Potassium	Electrolysis
	Sodium	
	Magnesium	
	Aluminium	
	Zinc	Carbon reduction
	Iron	
	Tin	
	Lead	
	Copper	Chemical reaction
	Silver	
Easy to extract	Gold	

Ecological , social and ethical issues associated with processing

Mining metal ores has harmful environmental effects:

- Large areas of natural countryside are ruined and left scarred and disfigured.
- Natural habitats of birds and wildlife are destroyed.

Mining and quarrying also has negative social effects:

- Dust can spread to surrounding residential areas.
- Noise from machinery and lorries at all hours can disturb local residents.
- Disused quarries can fill with water, which can be hazardous.
- Mine workings can collapse, causing subsidence under nearby buildings.

The extraction process also has harmful effects. Sulphur dioxide gas released during extraction reacts with rain and oxygen causing acid rain that can:

- damage forest ecosystems (trees, plants, animals, fish, soil and water supplies)
- cause breathing problems for humans
- corrode and seriously damage buildings and statues.

Lifecycle

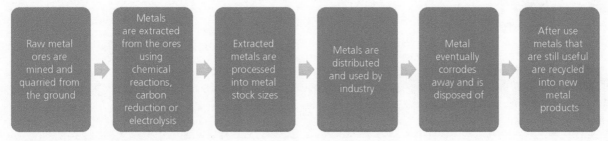

Figure 17.1 **The lifecycle of metals**

Recycling, reuse and disposal

- Almost all metals can be recycled over and over without loss of quality or strength.
- Scrap metal is collected in many ways and is an industry in itself, with a market value per tonne.
- Metal is sent to processing plants where it is separated into different metal types.
- The metals are then heated and melted back into liquid form using large furnaces called smelters.
- Once smelted, the metals are cast into small bars called ingots and are ready for reuse.

Commonly available forms and standard units of measurement

Sheet metal

- Most metals are available in sheet form.
- Sheet sizes range from around 500 × 500 mm to around 2.5 × 1.25 m, but many suppliers will supply sheets cut to a specific size.
- Common sheet thicknesses range from 0.025 mm up to 12.7 mm.
- Commonly used metals such as steel and aluminium are also available in perforated sheets with different mesh-like patterns.

Precious metals

Precious metals such as gold and silver can be bought in bars ranging from 1 g to 12.5 kg in weight, or in thin sheets called 'leaf' and thin wire strands for jewellery work.

Stock profiles

Commonly used metals are available in a wide range of stock profiles:

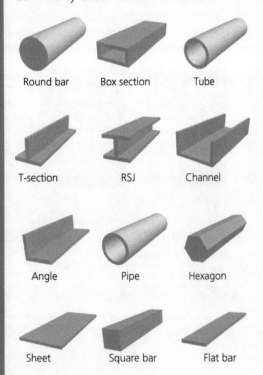

Round bar   Box section   Tube

T-section   RSJ   Channel

Angle   Pipe   Hexagon

Sheet   Square bar   Flat bar

Figure 17.2 Stock profiles

Standard components

Standard components are available for use when constructing and assembling products made from metal.

Nuts and bolts

Bolts are specified by:

- the type of head
- overall length
- material – usually steel or brass
- gauge – this relates to the bolt diameter
- finish – galvanised, zinc coated, chrome plated or japanned (a coloured lacquer).

Table 17.5 **Bolts and screws**

Bolt type	Illustration	Applications
Hex bolts		Used in machinery and construction. Can be fully or part threaded. Tightened with spanner or socket.
Machine screws		Used in machinery, construction and household products and appliances. Fully threaded. Tightened with a flat or crosshead screwdriver.
Thread-cutting machine screws		Used in construction, bodywork and steel-framed buildings. Point is 'self-tapping' so can be bolted into a non-threaded hole.
Sheet metal screws		Designed to be driven directly into sheet metal. Coarse thread 'grips' the hole in the sheet steel. Often called 'self-tappers'.
Socket screws		Similar to machine screws but with an internal hex socket for tightening with an Allen key.
Flange bolts		Used in engines and automotive applications. Bolt head has a flange that acts like a washer.

Pop rivets

- Pop rivets are cylindrical steel tubes used to join pieces of sheet metal.
- The rivet is inserted into holes in both sheets and a rivet gun pulls a pin through the cylinder, distorting and spreading the end of the tube so it blocks the hole and holds the two metal sheets together.

Hinges

For more information on hinges see Chapter 16.

Structural integrity

- Metal can be shaped, folded or curved in certain ways to improve its structural integrity.
- Flat steel bars will bend easily but by folding a flat steel bar into a right-angled section, its strength and rigidity are greatly increased.
- Sheet steel's strength and rigidity can be greatly improved by curving the metal.
- A tape measure's curved shape makes it stay rigid and not collapse when extended.

Figure 17.3 A flat and bent steel bar

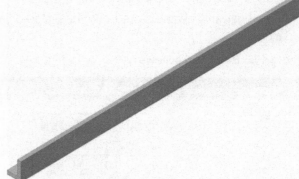

Figure 17.4 A folded steel bar

Heat treatments and processes

Various treatments and processes can be applied to metals to improve their structural integrity by making them harder, tougher or more ductile. Hardening and tempering increase the hardness and toughness of ferrous metals.

Hardening

- The hardening process is done first but makes the metal brittle.
- The metal is heated to around 1050 °C so it is glowing (red hot).
- It is then immersed in cool water, which cools it rapidly (called quenching).
- The process is repeated a number of times and the gradual heating/rapid cooling hardens the metal.

Tempering

- **Tempering** is done immediately after hardening to make it less brittle and likely to snap.
- The metal is heated gradually to 150–350 °C for around two hours and then left to cool gradually.
- Higher temperatures result in a softer but tougher metal; lower temperatures produce harder but more brittle metal.

> **Tempering:** A heating process applied to metal to make it less brittle.

Case hardening

- Case hardening is a simple process for hardening ferrous metals with a low carbon content.
- Case hardening only affects the outer surface of the metal so other properties are not altered.
- The metal is heated to around 1050 °C and then dipped into case hardening compound powder.
- The powder sticks to the hot surface of the metal, which is then re-heated before quenching to cool it rapidly.
- The process forms a tough outer layer on the metal and can be repeated to increase the hardness.

Annealing

- **Annealing** increases a metal's ductility and reduces its hardness, making it easier to work.
- The metal is heated gradually until it glows red, then allowed to cool in a very slow, controlled way.
- Annealing can also be done on non-ferrous metals such as copper, silver and brass.

> **Annealing:** A heating process applied to metal to increase its ductility and reduce hardness.

Finishes and surface treatments

- Some metals require no finishing of any kind as they have their own 'self finish'.
- Ferrous metals are very susceptible to corrosion and will begin to corrode almost immediately if no finish is applied.

Painting

- Surface rust, oil, grease and other substances are first cleaned from the metal.
- A primer is applied, which helps paint stick to the metal and stops rust reforming.
- Paint is then applied, by brush or spray.

Lacquering

- Lacquer can be applied by brush or spray application.
- Lacquer prevents tarnishing and oxidisation on the metal but dries clear, allowing the natural finish of the metal to show through.
- It is often used on non-ferrous metals, such as copper, brass and bronze, that have lustrous finishes to prevent tarnishing.

Stove enamelling

- Stove enamelling is similar to spray painting but the drying or 'curing' process is speeded up using heat.
- After spraying, the metal is baked at around 150 to 200 °C in an oven to give an extremely tough, durable finish.
- Stove enamelling is extremely heat resistant so is often used on cookers, wood-burning stoves and radiators.

Powder coating

- Powder coating is much more durable than paint or lacquer.
- The metal is first shot or sand blasted to remove oil, grease and any other material.
- The metal is then hung from a wire with a small electrical current applied to it.
- A dry powder made from polyester, polyurethane, epoxy or acrylic is then sprayed on to the metal.
- The spray gun gives the powder a positive charge so that it is attracted and sticks to the negatively charged metal.
- The metal is then baked in an oven at around 200 °C.
- The powder melts and forms a smooth, glossy, even coating over the metal which is extremely tough.
- Powder coating is used on bicycle and motorcycle frames, alloy wheels and household appliances such as washing machines and fridges.

Enamelling

- Enamelling involves melting glass powder on to metal, which forms a smooth, even coating.
- The glass powder is sprinkled on to the metal then heated to 750–850 °C; it then melts and spreads over the surface.
- Once hardened, it is smooth, durable, scratch and fade resistant, and has an easy-to-clean surface.
- A wide range of colours are available including clear and translucent finishes.
- Enamelling can crack or shatter if the metal it is applied to is bent or stressed.
- It is mainly used in jewellery and applied to gold, silver, copper and so on but it can also be used on stainless steel and cast iron.

Plating

Plating lays a thin layer of one type of metal on to another in order to:
- improve the appearance
- improve corrosion resistance
- improve hardness
- improve wearability
- reduce friction
- improve paint adhesion
- alter the metal's electrical conductivity.

Electroplating

- Electroplating is used to coat cheaper metals, such as copper, with a thin layer of more precious metals, such as gold or silver.

Electroless plating

- Electroless plating uses hydrogen; it reacts with metal ions to produce a negative charge and deposits metal ions on to the metal being plated.
- The electroless method deposits the metal more evenly over the object than electroplating so is often used for unusually shaped objects.

Galvanising

- Galvanising is a method of plating ferrous metals with a zinc coating to prevent corrosion.
- Before galvanising, the metal is washed in a series of solutions to remove all contaminates.
- The zinc is heated to 450°C until it melts, then the ferrous metal is completely immersed in the molten zinc.
- The zinc reacts with the ferrous metal, creating a series of metal alloy layers that bond across the metal surface.
- Galvanising is used on roofing sheets, metal buckets, industrial fencing and motorway Armco barriers.

Dip coating

- Dip coating coats the metal in a thin layer of plastic to prevent corrosion and improve grip.
- The metal is heated to between 250 and 400°C then powdered polymers such as polyethylene, nylon or PVC are applied to the metal.
- The powder melts and sticks to the hot metal, creating a smooth plastic coating over it.

Polishing

- Polishing and buffing enhance the natural 'self finish' of the metal.
- Polishing is more aggressive and uses progressively finer grades of abrasives to remove scratches and blemishes from the metal.
- Buffing is then used to give the metal a deep shine by applying polish by hand or using a buffing wheel.
- Different combinations of buffing mops and polishing compounds are used for specific metals.

Processes used to make early iterative models

REVISED

- Metal wire, mesh and rods can be used for developing ideas by making iterative models.
- Welding rods come in 900 mm lengths and can be bent into shape by hand or using pliers.
- Metal rods can also be used to create axles, pivots and other joints in working models or mechanisms.

- Thin metal sheets can be bent and formed by hand using hand tools and a vice.
- Steel or aluminium mesh such as chicken wire can cut, formed and bent to shape easily.
- Mesh can be used to form 3D models that can be covered with papier mâché and painted.
- Tin and aluminium foil can be scrunched up and moulded into shape by hand or used to cover Styrofoam or cardboard shapes to add strength or make them reflect light.

Manipulating and joining

Metal shears

- Metal shears or tin snips can used to cut straight or curved shapes out of sheet metal up to around 2 mm thick.
- Aviation shears work in the same way but are much easier to use due to an in-built mechanism that gives extra leverage.

Bench shears and throatless shears

- Bench shears work in the same way as tin snips but are larger and mounted on a bench.
- They have a long handle operating the blade that gives lots of leverage and can make straight cuts in sheets and flat bars up to 6 mm thick.
- Throatless shears are similar to bench shears but the two cutting blades separate completely, allowing metal to be moved around freely so that curved shapes can also be cut.

Nibblers

- Nibblers work in a similar way to tin snips and can be hand operated or powered by electric drills or compressed air.
- Nibblers cut away and remove a thin strip of metal from 3–6 mm wide called a 'kerf', which is thrown away.

Hacksaws

- Hacksaws have a G-shaped adjustable frame that holds the blade and keeps it under tension.
- Hacksaw blades have very fine teeth and can cut metal bars and rods.
- Junior hacksaws are smaller versions of the hacksaw with spring frames and thin blades for finer cuts in thinner metals.
- Panel hacksaws are gun-shaped holders that hold a hacksaw blade in such a way that it can be used to cut sheet metals.

Files

- Files are used to shape metal, smooth the edges of metal and for removing cut marks from edges.
- For more information about different types of files, see Chapter 16.

Drills

- Metals can be drilled to accept bolts, screws, rivets and so on with a hand or pillar drill.
- Standard drill bits for drilling softer metals such as copper and aluminium are made from high-speed steel (HSS).
- Harder metals such as stainless steel require cobalt or titanium carbide bits.
- Stepper drills have a stepped shape, allowing the drilling of progressively larger holes without changing the drill bit constantly.

Angle grinders

- Angle grinders are hand-held power tools with a spinning disc that can be used for cutting, grinding or sanding metal.

Oxyacetylene torches

- Oxyacetylene torches are used for cutting metals but are actually melting the metal away.
- Oxygen and acetylene gas are mixed and ignited, producing a flame of over 3500 °C from the torch.
- The torch heats metal until it is cherry red, then a trigger on the torch blasts oxygen on to the heated area.
- As the metal melts, it turns to liquid iron oxide and drips from the cutting area.

Plasma cutters

- Plasma cutters can cut metal extremely quickly and accurately by melting the metal in a fraction of a second.
- Inert gas is forced through the nozzle of the cutter on to the metal at extremely high speed.
- The gas has an electrical arc running through it, which turns to plasma when it touches the metal and melts it instantly.

Turning

- Metal lathes are used to 'turn' metal objects by spinning them and applying cutting tools to the surface.
- The cutting tools are mounted on a tool post that can move along the workpiece and in and out from the centre of rotation.
- Lathes can cut threads on to items and make components to high levels of accuracy.

Milling machines

- Milling machines have a rotary cutting tool with special teeth that spins at high speed.
- The piece of metal is clamped to a machine bed that can be moved in three different planes.
- The bed is raised up to the level of the cutter and along or across the cutter, which shaves off material.
- The rate it is moved against the tool (feed rate) and the cutter speed depends on the type of metal being machined.

Figure 17.5 A stepper drill

Gluing (bonding)

- Epoxy resins are strong adhesives made by mixing epoxy and hardener together in correct proportions.
- The epoxy and hardener react chemically to form an extremely strong adhesive that sets quickly once mixed.

Soldering

Soldering involves melting solder (a soft metal with a low melting point) on to metals to join them together.

Soft soldering

- Uses a soldering iron to heat the two metals; it is used mainly for joining wires and electronic components.
- Solder (a tin and lead or copper alloy) is fed into the joint in the form of a thin wire.
- The solder melts at 180–200 °C and flows around the joint then cools and fixes the joint together.
- Flux in the solder removes contaminates and helps it flow freely.

Silver soldering

- Uses a soldering torch powered by butane gas that produces a small, high-temperature flame.
- Used to join precious and semi-precious metals in jewellery making.
- Solder made from silver alloys is placed on the joint along with flux and the whole joint is then heated.
- The solder melts at 600–750 °C and flows around the joint.

Brazing

- Uses an oxyacetylene torch to join different metals together, such as bronze, steel, aluminium and copper.
- Flux is applied to the joint in the form of a paste before it is heated.
- The joint is then heated to over 1000 °C and the solder is fed into the joint in the form of a rod, which melts and flows around the two metals.
- Brazed joints are air- and water-tight, and can be plated over to give a seamless appearance.

> **Exam tip**
>
> Brazing is more difficult than welding but can be used to join different types of metal.

Welding

- Welding melts and fuses metals together and is the most effective way of permanently joining two metals.
  - Gas or torch welding is similar to brazing but uses higher temperatures so the two metals melt and fuse together rather than just the filler rod (solder).
  - Arc welding uses electricity to generate an electrical arc like a tiny lightning bolt between the two metals.
  - The arc generates such intense heat that it melts the two metals and filler metal, fusing them together.

> **Exam tip**
>
> Welding is only suitable on some metals.

Folding

- Metal can be folded by simply bending it in a vice and hammering it into a neat fold.
- Metal folders clamp and bend sheet metals easily and more accurately to form neat folds of up to 120 degrees.
- Box-and-pan brakes are special metal folders that can clamp and bend metal on more than one side to form box shapes.

Pressing

- Metal presses use hydraulic force to press sheet metal between two shaped dies.
- The press forms the sheet metal into the same shape as the surface of the dies without the use of heat.

Casting

- Metal is heated up to a molten liquid and poured into a mould.
- The metal cools and solidifies; the 'casting' is then removed from the mould.
- The rough casting is then filed and machined into its final shape.
- Casting is often used to create shapes that would be difficult to make using other methods.

Accurate marking out methods

REVISED

- Before cutting metal it must be marked out.
- Traditional methods use a dark blue dye called 'engineer's blue' and a scriber.
- Modern methods use thin permanent markers.
- When marking holes, a centre punch is used to mark the centre of the hole and provide a small indent for the drill bit to sit in to prevent it wandering.

Digital design tools

REVISED

- Many of the computer-aided manufacture (CAM) machines used for metals are computer-controlled versions of existing metal shaping tools:
  - CNC milling machines – the moving bed is controlled by computer.
  - CNC lathes – the movement of the tool and speed of the lathe is controlled by computer.
  - CNC plasma cutters – the movement of the plasma cutter across the sheet metal is controlled by a computer.
- Laser cutters are CAM machines used for engraving and cutting thin sheet metals.
- Laser cutters either cut all the way through or part-way through the material to a pre-set depth.
- Laser cutters can also cut polymers, card and timber.
- Small laser cutters are common in schools and can be easily programmed using simple CAD software.

Manufacturing at different scales of production and processes used

The scale of production and manufacturing method used to produce metal components and products depends on:

- Form – the shape of the parts, which limits the processes available.
- Budget – set-up costs can vary. High-volume manufacturing processes are expensive to set up but production costs are cheap and vice versa.
- Time – time taken to set up tooling and time to manufacture products.
- Material – material is chosen depending on form, function, availability and cost.

One-off production

- One-off production is used for making one or a small number of specialist products.
- It can be expensive and time consuming, and requires skilled workers to produce the required product.
- Some products are made specifically to a single client's requirements and are known as 'bespoke' items.

Batch production

- Batch production is used for making a set number of products that may have some variations.
- The product is made in stages by the same person or different people carrying out each stage.
- Batch production requires lower-skilled workers than one-off production.
- Items such as garden gates or hand rails, which may have a common shape but different variations, are often batch produced.

Mass production

- Mass production is used for making thousands of items or products that are exactly the same.
- Mass production often uses an assembly line where humans or machines carry out the same small task repeatedly on each part.
- Many of the processes, such as welding, riveting and positioning, are done by robots.
- Mass production is often continuous, with production lines rolling 24 hours a day and people working shifts.

Lean/just-in-time production

For information on lean/just-in-time production, see Chapter 20.

Cost and commercial viability, different stakeholder needs and marketability

- Metal prices can go up and down significantly, depending on supply and demand.
- Bulk discounting is often applied to metals, meaning the cost per tonne decreases as the quantity ordered increases.

- Large increases in metal prices will affect the cost of products made from metal.
- Costs can be reduced by:
  - using a cheaper metal or alloy with similar properties (for example pewter instead of silver)
  - plating cheaper metal with a more expensive metal so it looks the same (for example silver plating over copper)
  - choosing a lower grade of the same metal (using medium carbon steel instead of high carbon).

Calculating the quantities, cost and sizes

REVISED

- Stock metal sections are usually available in lengths of between 4 and 8 metres.
- Calculating the quantities required is important and should try to minimise off-cuts and waste.
- Waste and leftover metal pieces can be sold as scrap metal to recover some of the production costs.
- Scrap metal is recycled by specialist metal companies and reprocessed into new metal sections.

Example:

An outdoor bench has two metal end pieces made from 40 mm box section mild steel. Both ends of the bench are shown in Figure 17.9.

Figure 17.6 Diagram of a bench

Table 17.6 Cutting list

Piece	Number	Length	Total length
Top horizontal	2	400 mm	800 mm
Legs	4	300 mm	1,200 mm
Bracer	2	160 mm	320 mm
			TOTAL = 2,320 mm

40 × 40 mm mild steel box section costs £28.00 for a 4 m length.

To make the two bench ends you will need only one 4 m length but you will have 1.68 m waste.

To make two sets (four bench ends) you will need two 4 m lengths but will have enough left over to make another set of ends from the leftovers. Therefore it costs the same to make two sets as three sets.

Now test yourself

TESTED

1 Name an application of shape memory alloys (SMAs). [1 mark]
2 Explain the purpose of a blast furnace. [2 marks]
3 Describe two harmful environmental effects of metal extraction. [2 marks]
4 Name two tools that can be used to cut sheet metal. [2 marks]
5 Name three CNC machines used for shaping metal. [3 marks]

18 Thermo and thermosetting polymers

Physical and working properties

REVISED

- Manufactured plastics and natural materials such as rubber or cellulose are composed of very large molecules called polymers.
- Polymers are constructed from smaller molecular fragments, known as monomers, joined together.
- Natural polymers, including rubber and cellulose, have existed since ancient times.
- Synthetic polymers, also known as plastics, were developed in the early twentieth century.
- Synthetic polymers can be developed to have specific properties such as:
  - strength
  - stiffness
  - density
  - heat resistance
  - electrical conductivity.

Factors that influence selection

REVISED

- Synthetic polymers are generally easy to process and can be cost effective. This means high-quality products can be manufactured at relatively low cost.
- Polymers are generally:
  - lightweight
  - waterproof
  - tough
  - electrical and/or thermal insulators
  - resistant to atmospheric degradation (they will not rot like timber or rust like metal).
- With the addition of additives polymers can also be:
  - **transparent** or **opaque**.
  - **rigid** or **flexible**.

> **Transparent:** See-through.
>
> **Opaque:** Does not allow light to pass through.
>
> **Rigid:** Will not bend.
>
> **Flexible:** Bends easily without snapping.

Sources and origins

REVISED

Natural polymers

- Rubber latex is made from the resin from rubber trees and is used to make rubber bands and glues.
- Shellac is made from the lac bug and is used as a varnish to protect wood.

Synthetic polymers

- Vulcanite is made by adding sulphur to natural latex rubber and used to make combs and buttons.

- Bakelite, the first entirely synthetic polymer, was used to make radio cases and kitchenware.

Extraction and conversion

- Most modern polymers are entirely synthetic and made from carbon compounds obtained from crude oil.
- The crude oil is fractionally distilled to produce the chemical. For example, ethene is used to make the plastic polythene.
- An ethene molecule is made up of four hydrogen and two carbon atoms. Polythene is made by getting these to join together to form long chain molecules.

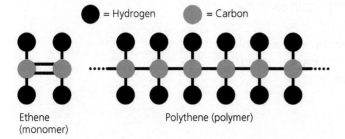

Figure 18.1 **Ethene molecules joined together to form polythene**

- Polythene molecules naturally join together to form high-density polythene. Once joined together, through the process known as polymerisation, they become known as a polymer.
- The mechanical properties of polymers can be improved by using additives such as:
  - **Plasticisers** – substances added to improve the flow properties for moulding. For example, PVC is used to make rigid drainpipes but, with the addition of a plasticiser, can be used to make a flexible garden hose.
  - **Pigments** – add colour to the polymer.
  - **Stabilisers** – help prevent damage from ultraviolet light, for example the polymer becoming brittle.
- **Fillers** add bulk to the polymer and reduce costs. They can also:
  - improve strength by reducing brittleness
  - increase resistance to impact.
- **Catalysts** can be used to speed up the synthesis of polymers, for example epoxy resin hardens in one hour compared to the usual 12.
- **Antioxidants** prevent oxidisation.

Ecological, social and ethical issues associated with processing

- World production of polymers has increased steadily to around 300 million tonnes a year.
- During the manufacture of polymers, significant quantities of **toxic** chemicals are produced. For example, manufacturing a polyethylene terephthalate (PET) bottle generates more than one hundred times the toxic emissions to the air and water than making the same bottle out of glass.

> Toxic: Harmful.

Safety concerns

- Producing plastics can be hazardous to workers, for example explosions, chemical spills or clouds of toxic vapour.

Environmental and health effects

- Many chemicals used in the production of plastics have negative environmental and human health effects. For example, phthalates are added to plastics to make them softer but have been classed as carcinogens (capable of causing cancer).
- The most obvious form of pollution associated with polymers is waste products that are dumped or sent to landfill.
- Around 10–20 million tonnes of plastic ends up in the ocean each year and this causes damage to marine ecosystems.
- It is estimated that every adult male in the UK generates one tonne of waste each year. Around nine per cent of this waste is polymer-based food packaging that could be recycled into new products.

Lifecycle

- Polymers are a very stable material and tend to stay in the environment for a long time after they are discarded, especially if shielded from sunlight in landfill sites.
- Polymers decompose very slowly, especially as most contain anti-oxidants to resist attack by chemicals.
- The average time for a PET bottle to degrade is 450 years and, during this time, significant quantities of toxic chemicals such as trichloroethane and methylene chloride are released.

Recycling, reuse and disposal

- Recycling recovers materials used in the home and industry for further use.
- Recycling can greatly contribute towards improving the environment by reducing the amount of material dumped in landfill sites.
- Many polymers can be recycled without loss of quality, for example a PET bottle can be recycled into clothing products such as fleeces.
- Polymer products that come into direct contact with food must be made from virgin polymer for health and safety reasons.
- The use of a certain amount of recycled polymer in a lower-quality product is called downcycling; for example, HDPE milk bottles can be used to create playground equipment.

Polymer recycling codes

Plastic products contain a symbol and number that are used to identify the type of polymer used to make the product. Identification of the specific polymer is essential for recycling.

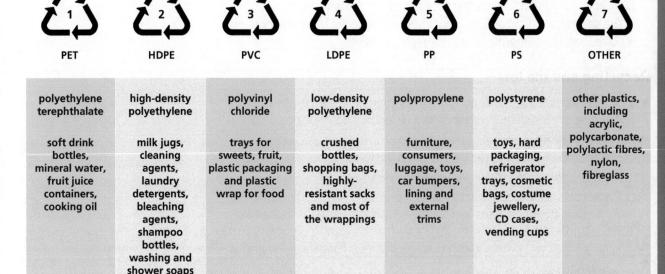

PET	HDPE	PVC	LDPE	PP	PS	OTHER
polyethylene terephthalate	high-density polyethylene	polyvinyl chloride	low-density polyethylene	polypropylene	polystyrene	other plastics, including acrylic, polycarbonate, polylactic fibres, nylon, fibreglass
soft drink bottles, mineral water, fruit juice containers, cooking oil	milk jugs, cleaning agents, laundry detergents, bleaching agents, shampoo bottles, washing and shower soaps	trays for sweets, fruit, plastic packaging and plastic wrap for food	crushed bottles, shopping bags, highly-resistant sacks and most of the wrappings	furniture, consumers, luggage, toys, car bumpers, lining and external trims	toys, hard packaging, refrigerator trays, cosmetic bags, costume jewellery, CD cases, vending cups	

Figure 18.2 Polymer recycling codes

Reuse

- Recycling is better than using new plastic but it does use a lot of energy for sorting, chipping and melting. Reusing a container or packaging is therefore better than recycling.
- Examples of reusing include:
  - refilling a bottle of mineral water with tap water
  - cleaning plastic jars and using them for storage
  - selling or giving away old toys.
  - adapting a product into something more useful, for example cutting down a plastic bottle to make a plant pot.

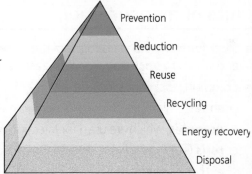

Figure 18.3 The three 'Rs' in the waste hierarchy

Reduce

- Consumers should be encouraged to buy products that use fewer polymers. For example, the new Nestlé Eco-Shape bottle uses 30 per cent less plastic.
- In 2014, 7.6 billion plastic bags were given to customers in the UK. This has been significantly reduced by shops now having to charge for plastic bags, encouraging shoppers to reuse bags or buy thicker, more usable 'bags for life'.

Reduce, reuse and recycle

- The slogan 'reduce, reuse and recycle' is a set of principles for the efficient use of resources. The best place to be is the top of the pyramid and the worst place to be is at the bottom of the pyramid. Prevention is, therefore, much better for the environment than disposal.

Disposal

- Environmentally friendly polymers are **biobased** or **biodegradable**.
- Polylactic acid (PLA) is a biobased polymer derived from cornstarch that breaks down into harmless chemicals when composted. It is used for disposable cups, cutlery and food containers.

> **Biobased:** A product from a renewable source.
>
> **Biodegradable:** The ability of a substance to break down naturally in the environment through the actions of micro-organisms.

● Biodegradable polymer products are made of normal (petrochemical) polymers but have additives that cause them to decay more rapidly in the presence of light and oxygen. They usually leave behind a toxic residue that makes them unsuitable for composting.

Recycling and the law

● Examples of laws and regulations that encourage manufacturers and consumers to consider the environment when buying or disposing of products include:
  ○ Electrical items, such as refrigerators, must have an energy rating label on them that allows consumers to make an informed choice.
  ○ Consumers are encouraged to take old products to collection points. As a result, designers have had to make products easier to dismantle, reuse and recycle.
  ○ New vehicles do not use toxic materials and all polymer parts must be labelled to ease recycling.

Commonly available forms and standard units of measurement

● Stock forms are standard forms of the material that can be purchased and stored in stock ready for manufacture; for example sheets of acrylic.
● Common stock forms include:
  ○ **sheet**, which can be cut and bent to make signs
  ○ **granules**, which are used for injection moulding
  ○ **rods** (solid), which are used for CNC machining
  ○ **extruded sections**, such as T, H and C sections, which are used to make ballpoint pens
  ○ **tubes** (hollow) in round, square or triangular section, which are used to make biros.
  ○ **foamed plastics**, such as expanded polystyrene, which are used to make models
  ○ **powdered polymers**, such as PVC, which are used to provide a coating on metal surfaces
  ○ **reels** of plastic wire that are used in some 3D printers.

> **Extruded:** A shape or material that has been produced by forcing it through a die.

Figure 18.4 Some common forms of polymer

Standard components

- A standard component is an individual part that is made many thousands of times to the same specification, for example a nylon washer.
- Standard components manufactured from polymers include:
  - **nuts and bolts** made to hold components together
  - **washers** used to secure fittings
  - **wall plugs** used to anchor screws into walls
  - **end caps** used to hide the heads of screws or close tubing
  - **gear wheels** used in toys and mechanical devices.

Figure 18.5 An assortment of polymer fixings

Structural integrity

Thermosetting polymers

- Thermosetting polymers are naturally hard and brittle but can be made stronger and tougher by using other materials to reinforce them, for example a fibreglass boat hull.

Glass-reinforced plastic

- Glass-reinforced plastic (GRP) is the reinforcement of polyester resin using strands of glass fibre available as a woven mat or loose strands.
- The glass-fibre reinforcement gives the material:
  - a much higher tensile and compressive strength
  - a lighter, hard-wearing surface
  - excellent resistantce to corrosion.

Thermopolymers

- Thermopolymers tend to be moulded as thin hollow shapes.
- The hollow shapes are designed with 'lips' and 'ridges' that make the shape more rigid and less likely to twist, flex and crack.
- Injection-moulded products often include stiffening 'ribs', rather than increasing the thickness of the entire wall of the product.
- Materials such as low-density polyethylene (LDPE) and polypropylene (PP) are attacked by ultraviolet light and, over time, may fade and become brittle. This is called UV degradation.
- To prevent UV degradation, manufacturers add UV stabilisers (similar to the chemicals used in suncream) which absorb the UV radiation.

Figure 18.6 A disposable cup with ridges and a rolled lip

Finishes and surface treatments

Industry finishes

- Polymers are usually described as self-finishing, which means they require no further finishing. For example, injection-moulded parts can be textured or highly polished but the finish is imparted to each product by the mould.
- Co-injection moulded products use two polymers; for example, a toothbrush handle may combine a hard polymer and a soft elastomer to improve grip.

Figure 18.7 A co-injection moulded toothbrush

- Once injection moulded, a product may be subject to one or more of the following finishing processes:
  - **Degating** – removing the runners and gates used to inject the polymer into the mould cavity.
  - **Deflashing** – removing excess polymer, called flashing, which has leaked out between the two halves of the mould.
  - **Cleaning** – removing the grease and dirt that can be picked up from the machine by spraying or dipping parts in a mild detergent.
  - **Decorating** – this may include printing and painting.

School-based finishes

School-based finishes include:

- Cleaning and polishing the edges of acrylic. The stages in this process are:
  - draw-filing the edges
  - moving through progressively higher grades of wet and dry paper
  - a final polish by hand or with a buffer.
- Trimming and finishing the edges of a vacuum forming. The stages in this process are:
  - trimming the edges with a craft knife or specialist trimming tool
  - scraping a steel ruler along the edges.

Processes used to make early iterative models

- Iterative modelling is repeated modelling to develop an idea. Polymers can be used in a variety of ways to develop iterative models.

Acrylic sheets

- Acrylic can be cut by hand but the edges take a long time to clean up. If this material is to be used for iterative modelling, then the rough edges may need to be left.
- It is better to use a laser cutter to cut acrylic for iterative models because:
  - The model can be made directly from the CAD drawing quickly and accurately.
  - An excellent edge finish will be achieved directly from the laser cutter.
  - Changes can be made to the CAD drawing quickly and the model made again.
  - Layers of acrylic can be built up to form complex models.
- A strip heater can be used to bend acrylic to make simple iterative models. For more complex shapes, the acrylic can be heated in an oven and formed around a simple wooden mould.
- Acrylic sheets can be glued together with dichloromethane (DCM) or another suitable polymer glue. A hot melt glue gun can also be used but the result will not look as clean or precise.
- Drilling can be used to create holes in acrylic but suitable plastic drill bits must be used.

Styrene sheets

- Styrene sheet is suitable for iterative models because it can be cut by scoring and then bending until it snaps along the line.
- Styrene sheet can be bent on a line bender.
- Styrene sheet can be glued together with dichloromethane (DCM) or another suitable polymer glue.
- Drilling can be used to create holes in styrene sheet but suitable plastic drill bits must be used.

Foam sheets

- Foam sheet, such as Plastazote, is available in a variety of thicknesses and can be cut with scissors or a craft knife.
- Foam sheets can be glued together with a cold melt glue gun or double-sided tape.

Sticky-backed vinyl

- Sticky-backed vinyl can be cut with scissors or a craft knife.
- More precise shapes can be cut out of sticky-backed vinyl using a CAD program and a CAM knife-cutting machine.
- Sticky-backed vinyl is available in a wide range of colours as well as mirrored and metallic finishes.
- Printable sticky-backed vinyl can be used to make stickers by printing a design from a computer directly on to the sheet.

Foam board

- Foam board is a thin sheet of expanded foam sandwiched between two layers of cardboard.
- It is usually 5 mm thick and is available in white or black sheets ranging from A4 to A1 in size.
- It is easily cut with a craft knife and safety ruler.
- It can be glued with an all-purpose adhesive, such as UHU, or a glue gun if time is an issue.
- It is often used for quick, box-like constructions such as architectural models.

Polystyrene foam

- Polystyrene foam, or 'blue' foam, can be used to make simple block models.
- It is often used to develop 'hands-on' models when exploring **ergonomic** suitability, for example the shape for a hairdryer.
- It is relatively cheap, easy to work, and comes in large, thick sheets.
- It can be cut with a band saw or simple hand tools and finished with an abrasive paper.
- Details can be added to polystyrene models using files, craft knives and marker pens.

> **Ergonomic:** Concerned with the physical shape of an object.

Acrylic shapes

- Modelling suppliers can provide a range a range of acrylic shapes that can be used for modelling. These include:
  - rod
  - tube
  - channels
  - hinges
  - nuts and bolts.

Modelling kits

- Model–making kits, such as Lego, were originally produced as toys.
- In recent years modelling kits, such as Lego Technic, have been developed that allow structures and mechanisms to be quickly modelled.

Manipulating and joining

Marking out

Before a design can be cut, it must be marked out. The following can be used to mark out on a piece of polymer:

- a chinagraph pencil
- a non-permanent marker (taking care as the lines can smudge)
- a permanent marker.

Wastage

Wastage is the process of cutting away material to leave the desired shape. It is called this because the material removed is thrown away (or recycled).

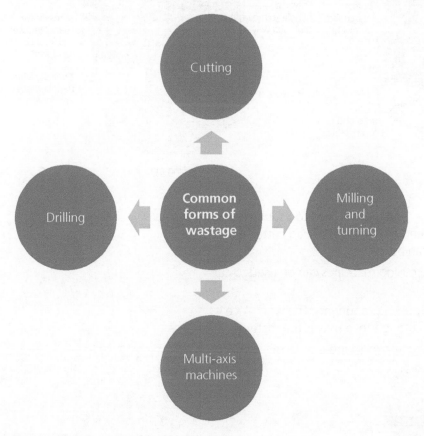

Figure 18.8 Common forms of wastage

Cutting

- Scissors can be used to cut thin foam sheets and 0.5 mm high-impact polystyrene (HIPS).
- Craft knives can be used to cut thicker polymer sheets by scoring and then flexing the material until it snaps along the line.
- A coping saw and scroll saw can be used to cut curves in thin polymer sheets. A junior hacksaw can be used to cut through sectional material such as rods.
- A band saw can be used to cut straight lines and shallow curves in larger sections of polymer sheet.
- A laser cutter can be used to cut 2D shapes in acrylic sheet but it is limited to the size of sheet the machine can take. A laser cutter can also be used to engrave designs into acrylic for decoration.

Milling and turning

- Milling machines can be used to cut slots and grooves in blocks of polymer. The two types of milling machine are vertical and horizontal.

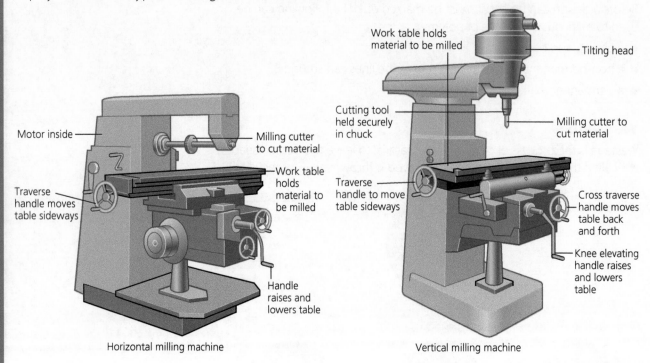

Horizontal milling machine

Vertical milling machine

Figure 18.9 Milling machines

- In both types of milling machine, the workpiece is fixed to a table that can move backwards, forwards, up and down under a fixed cutter. A variety of different shaped cutters can be used.
- CNC (computer numeric controlled) milling machines can be used to cut more complex shapes.
- Centre lathes can be used to make round components. The workpiece is held securely and rotates while the cutting tool removes the material.
- CNC lathes can be used to cut more complex shapes.

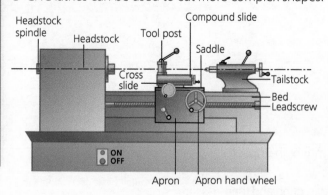

Figure 18.10 Centre lathe

Multi-axis machines

- Multi-axis machines can move in four or more ways in order to manufacture complex parts.
- A multi-axis machine can be considered a combination of a lathe and a milling machine, with the ability to turn a round shape and cut out slots.

Drilling

- Polymers can be drilled with hand and power drills but it is important to have the correct drill bit to make sure the material does not chip or crack.
- Drill bits for polymers generally have a shallower cutting angle and the drill speed is slowed down.
- When drilling a polymer, it is best to clamp it down securely with a piece of wood on the top and bottom to prevent the work moving and the underside chipping.

Addition

- Addition is the process of temporally or permanently joining materials together. Common addition methods for polymers include the following.

Adhesion

- Contact adhesive, such as Evo-Stik can be used to join thin polymer sheets to other materials.
- Epoxy resin can be used to glue some polymers together once the surfaces have been roughened.
- Polystyrene cement is used to join modelling kits, such as Airfix kits.
- Tensol is a solvent-based adhesive used to join acrylic.
- DCM is a thinner version of Tensol that can be used to join acrylic and HIPS.
- Double-sided tape can be used as a semi-permanent method of joining polymers together.

Heat welding

- Heat welding can be carried out on polymers using a hot air welding tool.
- The hot air welding tool heats up the material to be joined and feeds a thin wire polymer into the area to be joined.
- Polymers can also be joined by ultrasonic welding, which uses high-frequency electromagnetic waves to soften the polymers.

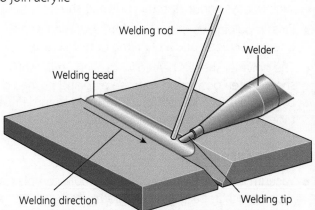

Figure 18.11 Heat welding plastic

Mechanical fixings

Polymers can be joined using a wide range of mechanical fixings, for example machine screws, bolts, self-tapping screws, rivets and clips.

Deforming and reforming

Common deforming and reforming methods for polymers include the following.

Line bending

- Line bending is a process used to create simple bends in polymer sheets such as acrylic.
- The sheet material is heated along a line with a strip heater until it softens. It is then folded and held in place until it cools and hardens.

Drape forming

- Drape forming is used if a large curve or bend is required.
- The polymer sheet is heated in an oven until it softens. It is then draped over a former and a piece of cloth is pulled tightly across it to hold it in shape until the polymer has cooled.

Press moulding

- Press moulding is used to produce more complex shapes such as trays or dishes in thin polymer sheet.
- Press moulding uses two-part formers called a yoke (the upper piece) and a plug (the bottom piece).
- The polymer sheet is heated in an oven until it softens. The sheet is then positioned over the plug and pushed down by the yoke. Both parts of the former are then clamped together until the material has cooled.

Vacuum forming

- Vacuum forming is used to produce more complex large shapes, such as packaging and baths, from polymer sheets.
- The polymer sheet is clamped in position and heated until it softens. Below the plastic sheet is a mould that is raised and then the air between the mould and the plastic sheet is sucked out by a pump. Once the material has cooled, it is removed from the machine.

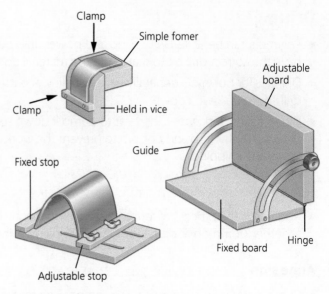

Figure 18.12 **Line bending**

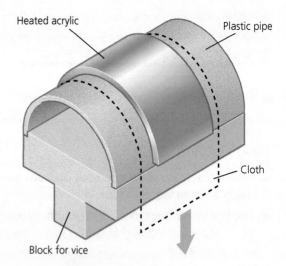

Figure 18.13 **Drape forming**

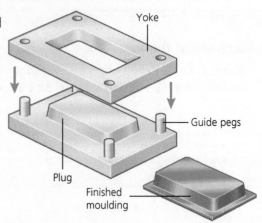

Figure 18.14 **Press moulding**

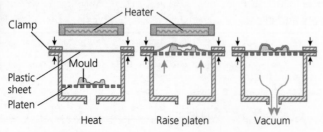

Figure 18.15 **The vacuum forming process**

- The sides of a vacuum forming mould must be tapered and have radiused (rounded off) corners to allow the forming to be released from the mould.

Digital design tools

- Industry professionals use a range of digital tools when exploring and developing design ideas. These include rapid prototyping and digital manufacture.

Rapid prototyping

- Rapid prototyping is making 3D solid objects from a digital file.
- The object is made using additive processes, where the object is created by building up successive layers.

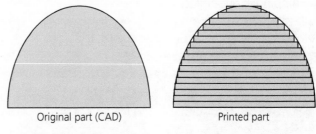

Original part (CAD)   Printed part

Figure 18.16 Rapid prototyping builds up objects in layers

Digital manufacture

Computer-aided manufacture (CAM) can be used in the following ways:

Stereo lithography

- The component is built up in layers of liquid photopolymer. When the liquid photopolymer is exposed to ultraviolet light from the laser beam, it cures or solidifies. The platform moves down and the process is repeated.

Laser sintering

- In this process, powder is spread over a platform by a roller and then a laser sinters selected areas, which makes the powdered polymer melt and then harden.

Fused deposition modelling (FDM)

- In this process, a polymer filament is passed through a heating element, melted and extruded. Each slice of the model is drawn from a continuous length of the molten filament.
- Typical build up materials are acrylonitrile-butadiene-styrene (ABS) and polylactic acid (PLA).

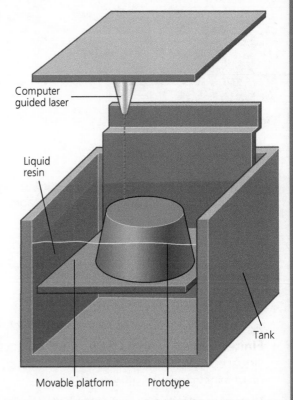

Computer guided laser

Liquid resin

Tank

Movable platform   Prototype

Figure 18.17 Stereo lithography

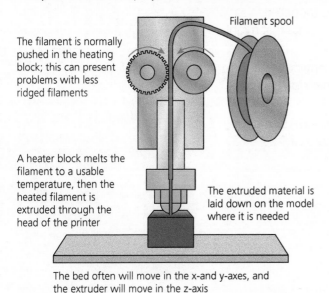

Filament spool

The filament is normally pushed in the heating block; this can present problems with less ridged filaments

A heater block melts the filament to a usable temperature, then the heated filament is extruded through the head of the printer

The extruded material is laid down on the model where it is needed

The bed often will move in the x-and y-axes, and the extruder will move in the z-axis

Figure 18.18 Fused deposition modelling (FDM)

18 Thermo and thermosetting polymers

3D printing

- In this process a thin layer of build material, typically plaster or starch based, is gradually glued together by a print head that can also colour the material.
- The final product is cleaned and the outer surface strengthened by dipping in Superglue.

Figure 18.19 A 3D printer

Digital technologies

- **Computer-aided design (CAD)** is used to create and explore ideas through 2D and 3D drawings. The advantages of using CAD are:
  - ○ changes can be made quickly
  - ○ drawings can be rendered in different colours and materials
  - ○ components from built-in or online libraries can be used
  - ○ designs can be shared electronically.
- **Computer-aided engineering (CAE)** is the broad use of computer software to aid design engineers in analysis of tasks.
- **Finite element analysis (FEA)** allows products to be modelled in a virtual environment and tested for weaknesses, such as areas that might need reinforcing with stiffening ribs.
- **Virtual modelling** is used to see how a liquid polymer would flow into and around a mould cavity. Such factors as polymer injection pressure and cooling time can be considered.
- **Draft analysis** is concerned with calculating the optimum draft angle required for a product to be easily released from a mould cavity.

Manufacturing at different scales of production and processes used

Which scale of production is used for polymer parts depends upon:

- **Form** – what shape the parts are.
- **Budget** – how much money has been allocated to produce the parts.
- **Time** – how much time has been allocated to make the parts.
- **Material** – which polymer is to be used.

One-off/low-volume polymer production

One-off production involves making only one or a small number of products. One-off production techniques include:

- **GRP layup** – a technique applying liquid polyester resin and glass fibre mat or woven fabric to a metal, wood or plaster mould.
- **GRP moulding** – pressure is applied to the top surface of the moulding of resin/catalyst/glass fibre. By heating, comparatively fast hardening of the resin is possible.
- **Fabricating** – parts are made by gluing, turning, carving or welding existing materials.
- **CNC machining** – an example of subtractive manufacturing, which means material is cut away from the block.
- **3D printing** – a thin layer of build material, typically plaster or starch based, is gradually glued together by a print head that can also colour the material.

1. Polish mould   2. Brush on release agent   3. Brush on gel coat 5 mm thick

Figure 18.20 GRP layup

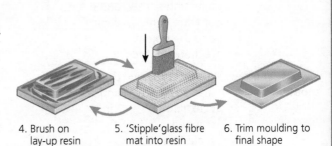

4. Brush on lay-up resin   5. 'Stipple' glass fibre mat into resin   6. Trim moulding to final shape

Figure 18.21 GRP moulding

Batch/medium-volume polymer production methods

Batch production involves making a set number of identical products. Batch production techniques include the following.

- **Vacuum forming** – you can read about vacuum forming earlier in this chapter.
- **Casting** – this involves pouring liquid resin into a mould that then solidifies. The moulds are cheap and can be reused a number of times.
- **Rotational moulding** – in this process a heated mould containing plastic paste or powder is rotated. As the mould rotates, the plastic evenly coats the mould cavity. Products made by rotational moulding include tanks, barrels and large hollow toys.

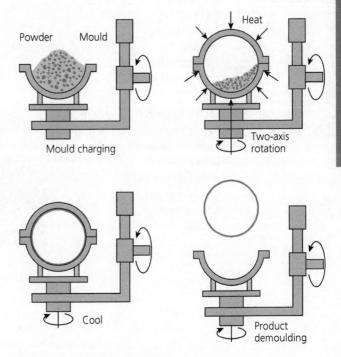

Figure 18.22 Rotational moulding

Mass/high-volume polymer production methods

Mass/high-volume production involves producing a large number of products.

Table 18.1 Methods of mass volume production

Method	How it works	Products used for	
Compression moulding	Shaping thermosetting plastics by heating and compressing them into shape	Electric plugs, sockets and switches	
Injection moulding	Molten plastic is injected into a mould via an injection screw or ram, cooled, and the object is then ejected	Buckets, gear wheels and plastic chairs	
Extrusion blow moulding	A short tube of melted plastic is extruded, trapped in a mould and then air blown in so that the plastic takes the shape of the mould cavity	Bottles and petrol tanks	
Thermoforming, or sheet moulding	A continuous extruded sheet is heated, passed over a mould, and shaped by pressure and/or a vacuum	Vending cups and yoghurt pots	
Continuous extrusion	Molten plastic is pushed continuously through a shaped hole (profile or die) before being cooled	Garden hose and guttering	

Method	How it works	Products used for	
Calendering	Passing thermoplastic compositions through heated rollers to produce thin sheet	PVC flexible film and PS rigid foils	
Tubular sheet or sheath extrusion	Extruded plastic is expanded into a sheath and then wound on to reals	Packaging films and plastic bags	

Cost and commercial viability, different stakeholder needs and marketability

REVISED

The choice of polymers for production depends upon cost and availability.

Raw materials

- The cost of the raw material for most polymers is the same as it is linked to the price of crude oil.
- High-performance polymers cost more and include:
  - polyacetals – used in medical devices
  - Kevlar – a fibre used in bullet-proof vests
  - polyether ether ketone (PEEK) used in the aerospace industry.
- Costs can be reduced by using recycled plastic, but the costs of recycling can be high.

Responding to stakeholder needs

A key strength of using a polymer for a product is the ability to fine-tune its properties to meet stakeholder needs by introducing additives such as:

- **plasticisers** to improve the flow of the polymer into a mould and reduce production times
- **pigments** to give the customer a wide choice of colours.

Manufacturing

The advantages and disadvantages of common manufacturing methods include:

- **Injection moulding** – requires a complex mould but parts are produced at high speed and the cost of the mould can be spread over the production run.

- **Blow moulding** – simpler than injection moulding and generally less expensive.
- **Thermoforming** – offers lower tooling costs and is used for lower-volume production runs.

Calculating the quantities, cost and sizes

Quantity

- The amount of material used to make a product is referred to as the quantity of material required. For example, a plaque might be cut from a 100 mm × 50 mm rectangle of 3 mm thick acrylic sheet.
- If more than one product is to be made, the quantity increases; for example, ten plaques could be cut from a 200 mm × 250 mm rectangle of 3 mm thick acrylic sheet.

Cost

- If a 1000 mm × 500 mm sheet of 3 mm acrylic sheet costs £10.00, then the material costs for one plaque is £0.10 and for ten plaques £1.00.
- In industry the calculation of the costs includes:
  - **Fixed costs** – the cost of making tools and fitting them into the machine.
  - **Variable costs** – the costs of making of parts, including raw materials, labour and energy.
- The cost of making goes down as the number of parts made increases.

Number of parts made	1,000	2,000	5,000	10,000
Fixed cost (£)	2,000	2,000	2,000	2,000
Variable cost (£) @ 50p per part	500	1,000	2,500	5,000
Total cost (£)	2,500	3,000	4,500	7,000
Cost per part (£)	2.50	1.50	0.90	0.70

Table 18.2 The cost of making products

- Injection moulded parts can be carefully designed using CAD and changes made to reduce costs, for example the number of components formed in a single injection.

Sizes

- All polymers that are moulded by heat will shrink as the moulding cools. The finished size of the product must be kept within tolerance (the allowable amount of variation in size).
- Shrinkage is not always easy to predict and depends upon the following factors:
  - type of polymer used
  - thickness of the walls
  - mould temperature
  - pressure used in the moulding.

Tolerances and minimising waste

Lean manufacturing principles are applied to polymer manufacturing to reduce the amount of waste. Key factors include:

- Designing parts correctly to avoid costly and time-consuming quality issues during manufacture.
- Minimising material by judging the correct wall thickness (a balance between reducing costs and maintaining strength).
- Reducing the cycle time by designing moulds that allow parts to cool rapidly and be ejected.
- Reusing waste material by feeding excess material back into the moulding machines.

Exam tip

In preparation for the exam, make sure you develop a clear picture in your mind of:
1 The working properties of a range of polymers.
2 The manufacturing process used to make a range of polymer products.

Typical mistake

Students often fail to fully justify their answers in exams. For example, they may state that low-density polythene (LDPE) is a good material for carrier bags but fail to state the working properties (range of colours, tough, flexible, and so on) that make it a good choice.

Now test yourself

TESTED

1 Name two general properties of polymers.	[2 marks]
2 Name two additives that are added to polymers.	[2 marks]
3 Name two polymer stock forms.	[2 marks]
4 What is meant by the term UV degradation?	[2 marks]
5 Name two manufacturing methods used in one-off production.	[2 marks]

19 Fibres and textiles

Physical and working properties

A designer needs to consider the physical and working properties of fibres and fabrics carefully before selecting an appropriate material to use, as a fibre's properties will affect its suitability for the functionality of a product.

Different fibres have their own unique physical and working characteristics, but in general:

- natural fibres, such as wool, silk and cotton, are absorbent
- synthetic fibres, such as polyester and nylon, have a diverse range of properties and are very versatile.

Fibres can be mixed or blended together to benefit from the properties of more than one fibre – improving the function, aesthetic value or cost of the final product.

Typical mistake

Make sure you know and understand some of the different types of fibres. In an exam you should be able to select and explain the properties of both natural and synthetic fibres. Many candidates get confused and lose marks by incorrectly labelling fibres.

Natural fibres

Table 19.1 **The physical and working properties of a range of natural fibres**

Fibre	Properties/working characteristics	Uses
Cotton	Absorbent, strong, good handle, cool to wear, drapes well, creases easily, easy to handle and sew, will not stretch, burns easily	Medical dressings, T-shirts, socks, denim jeans, cosmetic pads, nappies, bed sheets, upholstery, canvas, car tyre cords, fishing nets
Linen (flax)	Absorbent, strong, good handle, cool to wear, drapes well, natural lustre (shine), creases easily, easy to handle and sew, will not stretch, burns easily	Bed sheets, table coverings, tents, skirts, suits, upholstery, canvas, wallpaper, bank notes
Wool	Absorbent, strong, good handle, warm, crease resistant, shrinks easily, drapes well, difficult to handle and sew, stretches easily, some flame resistance	Coats, jumpers, sportswear, blankets, socks, insulation, soundproofing, snooker tables, carpeting
Silk	Absorbent, strong, good handle, good insulator (cool in summer and warm in winter), drapes well, high natural lustre (shine), crease resistant, difficult to handle and sew, low stretch (can be stretched out of shape), burns slowly	Evening wear, ties, handkerchiefs, bed sheets, medical dressings, parachutes, sutures (stitches), wall coverings

Synthetic fibres

Table 19.2 The physical and working properties of a range of synthetic fibres

Fibre	Properties/working characteristics	Uses
Polyester	Non-absorbent, strong, good handle, poor insulator, durable, crease resistant, will not stretch, melts easily	Clothing, pillow filling, upholstery padding, bedding, carpeting, thread, ropes, boat sails
Nylon (polyamide)	Non-absorbent, abrasion resistant, very strong, some elasticity, durable, Resistant to chemicals and perspiration	Seat belts, tents, parachutes, rucksacks, shoelaces, toothbrush bristles, umbrellas, life jackets, tights, underwear, carpeting
Acrylic	Water resistant, quick drying, strong, good insulator, resistant to chemicals and perspiration	Fleece, ski jackets, blankets, rugs, outdoor furniture, knitwear, cleaning cloths
Viscose (rayon)	Absorbent, good insulator, creases easily, will not stretch, weak fibre, particularly when wet	Blouses, sportswear, shirts, blankets, curtains, table cloths, upholstery
Elastane	Non-absorbent, excellent elasticity, resistant to chemicals and perspiration, quick drying	Swimwear, sportswear, denim jeans, leggings, tights

Factors that influence selection

REVISED

When selecting materials and components, as well as considering their physical properties and working characteristics, a designer also needs to consider the following factors:

- Functionality needed: will it do the job it needs to do? Is it a suitable material?
- Aesthetics: will the material give the right look and texture to the product?
- Environmental considerations: does the material cause any harm to the environment throughout its lifecycle?
- Availability and cost of stock forms: is the material affordable within the budget set for the product?
- Social, cultural and ethical considerations: have all potential issues surrounding the material's lifecycle been addressed?

Sources and origins

Origins of natural fibres

Natural fibres can be classified into two categories: **natural protein fibres** and **natural cellulose fibres**.

> **Natural cellulose fibres:** Fibres that come from plant-based sources, for example cotton and linen.
>
> **Natural protein fibres:** Fibres that come from animal-based sources. These include hair, fur or silk fibres.

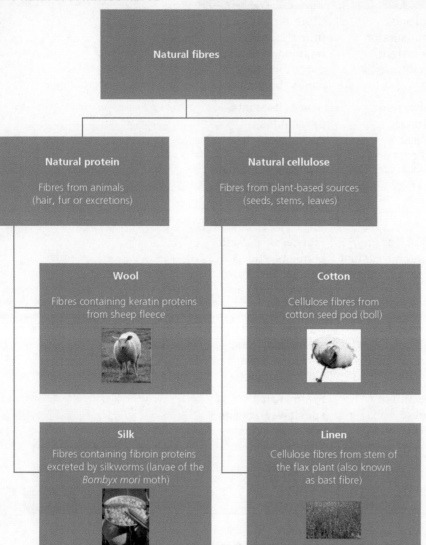

Figure 19.1 Origins of natural fibres

Table 19.3 **Leading producers of natural fibres**

Fibre	Leading producers
Cotton	Brazil, Pakistan, Turkey, USA
Linen	Canada, Russia, France (the plant flourishes in cooler climates)
Wool	Australia, New Zealand, USA
Silk	China

Origins of synthetic fibres

- Synthetic fibres are extracted from petrochemicals such as coal and oil.
- They are made up of polymer chains formed when single molecules called hydrocarbon monomers bond together in a process called polymerisation. **Natural polymers** (cellulose) can be chemically extracted from plant-based sources and processed to form a different type of synthetic fibre, known as a Negenerated fibre.

> **Natural polymers:** Chains of protein or cellulose molecules (monomers), such as keratin, glucose or fibroin. These chains are the basis of all natural fibres.

Now test yourself answers and quick quizzes at **www.hoddereducation.co.uk/myrevisionnotes**

Extraction and conversion

Natural fibres

Cotton

- Cotton is intensively farmed across huge areas of land using fertilisers, pesticides and large amounts of water.
- Fibres are harvested from the seed pod (boll) of the cotton plant.
- Cotton must be cleaned and bleached before it can be spun into yarn for fabric manufacture.
- A cotton plant produces around a hundred cotton bolls annually – enough to produce 225 pairs of jeans.

Figure 19.2 Thousands of hectares of land are used for cotton farming

Linen

- Linen is harvested from the flax plant.
- Flax stems are left to dry and the tough, outer layer rots away. This process is called **retting**.
- The stems are then broken and the fibres removed from inside.
- The fibres are cleaned and prepared for manufacture into yarn.

Wool

- Wool is sheared from the sheep before being sent to a mill.
- Wool contains a natural oil called lanolin, which makes it greasy. A process called scouring (which uses warm water and detergents) is used to remove this grease before the fibres can be used.

Figure 19.3 Raw linen fibres must be treated before use

Silk

- Silkworms are bred in captivity and moult as they grow. After the fourth moult, the worm encases itself in a cocoon of silk fibre and begins to pupate.
- The cocoons are boiled to kill the pupa inside and to remove the seracin (a gum that helps the silk fibres stick together when producing the cocoon).
- Silk manufacture is labour intensive, which makes silk expensive.

Synthetic fibres

Synthetic fibres are produced using extrusion processes:

- Polymer solution or melted polymer pellets are placed in a tank above a spinneret.
- The solution is forced through the tiny holes in the spinneret using air pressure, or a syringe-like system, forming the solution into long individual fibres.

Types of fibres

Three types of fibre are used in fabric and yarn production:

- **Staple fibres** are short fibres with a crimp (wavy texture) that can be spun into yarns, felted or bonded. All natural fibres, apart from silk, are staple fibres.

- **Filament fibres** are long, smooth fibres. All synthetic fibres and silk are filament fibres. These do not mesh well and therefore are cut short and heated to give them the appearance of a staple fibre.
- **Microfibres** are tiny filament fibres made from synthetic polymers such as polyester or nylon. They are very lightweight and versatile, and are used in clothing, cleaning cloths and insulation.

Converting fibres into yarn

- Staple fibres are carded (combed) so that they lie in the same direction before they are spun into yarns.
- Spinning involves pulling and twisting the staple fibres so that they mesh together.

Woven fabric construction

- Woven fabrics are mostly produced on industrial machinery using warp and weft yarns (although they can be produced on traditional wooden looms).
- Warp yarns run vertically down the length of the fabric (this is called the straight **grain**).
- The selvedge is the edge of the fabric roll. It runs parallel to the straight grain and prevents the fabric from fraying or unravelling.
- Weft yarns run horizontally across the fabric; this is called the cross grain.

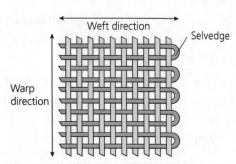

Figure 19.4 Warp, weft and selvedge of fabric

Table 19.4 **Types of weave**

Weave	Characteristics	Examples	Illustration
Plain weave	• The simplest and most widely used weave structure. • Versatile – different weights of fabric can be produced by altering the spacing of the warp and weft yarns or by using coarse or fine yarns.	Muslin (a lightweight plain weave fabric) Calico (a medium-to-heavyweight plain weave fabric)	
Twill weave	• Easily recognisable by the diagonal pattern formed by the crossing of the warp and weft yarns. • Produces a heavier fabric that is stronger and more durable than a plain weave. • Decorative (because of its diagonal pattern).	Denim Canvas Herringbone and houndstooth fabrics	
Satin weave	• Produces a smooth, lustrous finish due to the 'floating yarns' on the surface of the fabric – the weft yarns are woven under one warp yarn, then over a minimum of three warp yarns. • Has a shiny finish because of its larger surface area, which reflects light. • Weak and easily snagged.	Brocade, brocatelle, crêpe-satin	

Non-woven fabrics

- Non-woven fabrics are created using the shape and texture of staple fibres. They are not woven or knitted from yarns.
- Types of non-woven fabric include bonded fabrics and felted fabrics.
  - **Bonded fabrics** are manufactured by applying pressure, heat or adhesives to synthetic fibres to bond them together. They lose their strength and structure once wet so are usually used in disposable items such as wet wipes, tea bags, surgical dressings and nappies.
  - **Felted fabrics** are produced by applying moisture, heat and friction to matt staple fibres (usually wool and acrylic) together. Felt is often used for decorative purposes (such as appliqué) on the surface of pool and snooker tables, and for cushioning and insulating. It is a weak fabric that loses its shape when wet.

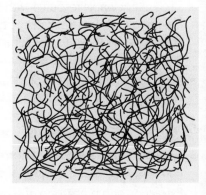

Figure 19.5 A non-woven fabric structure

Knitted fabrics

- Knitted fabrics are loose, flexible fabrics in which yarns are looped together in rows of interlocking loops (called stitches).
- The most common knitted fabrics are weft knit and warp knit.

Table 19.5 Weft knit and warp knit fabrics

Weft knit fabrics	Warp knit fabrics
A single continuous yarn constructed in horizontal rows of interlocked loops: • The horizontal rows are called courses. • The vertical columns are called wales.	Comprised of multiple yarns in vertical rows of interlocking loops that zigzag from side to side.
May snag and can unravel if the yarn is damaged or pulled.	Flexible; does not run or unravel.
Often produced using knitting needles (although they can also be manufactured on automated knitting machines).	Can only be completed on automated machines.
Used for a wide range of knitwear and home furnishings.	Suitable for sportswear and swimwear.

Ecological, social and ethical issues associated with processing

Ecological issues

Farming

- Natural fibre crops such as cotton and linen are intensively farmed on vast areas of land using large amounts of water.
- Toxic pesticides and fertilisers can contaminate soil and water supplies – killing wildlife, contaminating drinking water and removing nutrients from the soil, which reduces the quality of future crops.
- While organic farming uses natural fertilisers and pesticides, it still uses large amounts of water.
- Livestock farming for wool production has less of an ecological impact than crop farming, but land can be damaged by overgrazing; chemical dips containing insecticides can contaminate water and soil.

Figure 19.6 Toxic pesticides are used heavily on fibre crops

- Livestock also produce large amounts of the greenhouse gas methane.
- Silk is a renewable resource with little ecological impact, but the breeding of silkworms in captivity is heavily criticised by animal welfare activists as it has led to the demise of the *Bombyx mori* moth in the wild, and captive moths have evolved to be blind and unable to fly.
- More sustainable sources of natural fibres include soured milk, fermented wine, spider silk, corn husks and lab-grown bacteria.

Synthetic fibre processing

- Synthetic fibres are sourced from finite resources that are not sustainable and cause significant ecological damage.
- Extraction and processing of synthetic fibres uses high levels of energy and chemicals, and produces contaminated waste.
- Processing plants pollute water and emit greenhouse gases that contribute to air pollution.
- Synthetic fibres are non-biodegradable and take hundreds of years to decompose.

Transportation

- Transporting raw fibres causes air pollution and uses a significant amount of non-renewable resources.

Washing, dyeing and printing

- Washing, drying, dyeing and printing of fabrics requires lots of energy.
- It generates large amounts of waste water, which can contaminate local water sources and damage habitats.

Packaging

- Over-packaging of textile products creates large amounts of plastic waste.

Social and ethical issues

- Garment workers in some countries are subjected to poor and dangerous working conditions, long hours, low pay and little protection from union rights.
- Child labour, although illegal, is common in some countries.
- There are a number of work-related deaths in the textile industry each year.
- Demand for **fast fashion** has led to a **throwaway culture** in which cheap clothing is worn for a short period of time before being thrown away. Low prices mean some manufacturers have reduced workers' pay and conditions to maintain profits.
- The health of people who live and work near textile industry factories may also be affected by the contamination of land and water supplies, and exposure to dyes and toxic chemicals.
- Textile manufacturers have a responsibility to consider and protect the health and safety of their workers, their families and those living nearby.

Fast fashion: A recent trend involving the quick transfer of new collections from the catwalk into shops. Fast fashion is often on-trend, low quality and low in price. The characteristics of fast fashion mean that consumers buy large volumes of clothes more regularly, creating more profit.

Throwaway culture: The rise of fast fashion has made clothing more affordable for consumers. The incredibly low prices in some high street stores have resulted in a throwaway culture, meaning that consumers don't feel the need to keep clothing that is no longer in fashion, and happily dispose of it.

Lifecycle

- Natural fibres are biodegradable and decompose naturally in the environment within six months. They release carbon dioxide during decomposition, as well as dyes and chemicals.
- Synthetic, polymer-based fibres decompose very slowly – they take between 450 and 1000 years to break down – and release toxic and greenhouse gases into the atmosphere.

Recycling, reuse and disposal

- Most fibres and fabrics can be easily recycled.
- Clothing banks and charity shops ensure that waste textiles are reused or recycled effectively.

Figure 19.7 Employees work for long hours for little money, and often in poor working conditions

The three types of recycling are outlined in Figure 19.8.

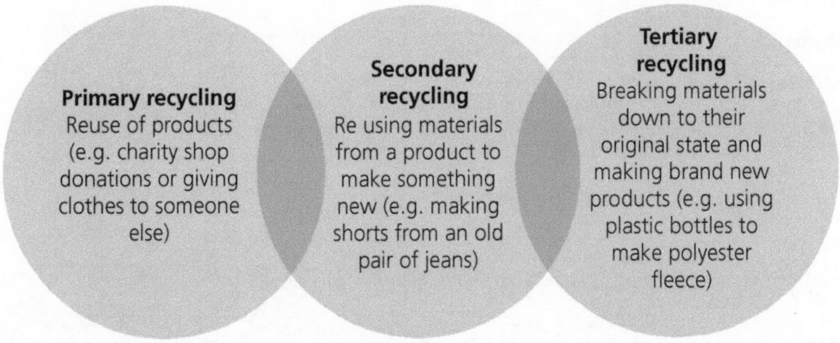

Primary recycling
Reuse of products (e.g. charity shop donations or giving clothes to someone else)

Secondary recycling
Re using materials from a product to make something new (e.g. making shorts from an old pair of jeans)

Tertiary recycling
Breaking materials down to their original state and making brand new products (e.g. using plastic bottles to make polyester fleece)

Figure 19.8 **The three types of recycling**

Commonly available forms and standard units of measurement

Stock forms

- Fabrics are usually bought 'off the roll' in standard widths of 90 cm (interfacings or linings), 115 cm and 150 cm.
- Fat quarters are pre-cut cotton fabrics used for smaller projects. They are available in bundles of complementary colours and patterns.

Figure 19.9 Fabrics are usually bought 'off the roll'

Table 19.6 Common fabrics and their uses

Fabric name	Example use
Drill	Upholstery
Jersey	T-shirts
Denim	Jeans
Voile	Mosquito nets
Tweed	Jackets
Gabardine	Suiting
Felt	Appliqué, crafts
Broadcloth	Dresses
Sheeting	Bed sheets
Damask	Decorative napkins
Muslin	Sheer curtains
Satin	Occasion dresses
Crêpe	Blouses
Velvet	Eveningwear
Corduroy	Trousers
Lace	Decorative
Chiffon	Lingerie
Organza	Drapes

Standard components

Threads

Threads come in various types, weights and textures.

Table 19.7 Common threads and their uses

Thread	Description	Use
Machine thread	Strong thread	Machine sewing
Tacking thread	Strong, thick thread	Hand-sewing seams before machine sewing
Embroidery thread	Thick, lustrous thread	Decorative work
Monofilament	Very strong, usually clear, thread	Invisible stitching

Fastenings

Fastenings are selected depending on their aesthetic, function and intended use.

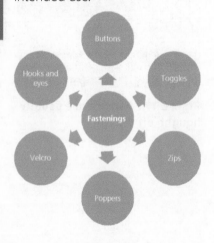

Figure 19.10 Fastenings

Figure 19.11 Hook and eyes are a common fastening used in bras

Now test yourself answers and quick quizzes at www.hoddereducation.co.uk/myrevisionnotes

Structural components

Structural components add support to a fabric or help shape a garment.

- Boning is a metal or plastic strip that is sewn into reinforced seams to add structure and shape to a garment, such as a corset.
- Petersham is a heavyweight ribbon or band used to stiffen waistbands or reinforce button bands.
- Interfacing is a light- to medium-weight bonded fabric used to stiffen collars, cuffs or armholes.

Other components

- Bias binding is a strip of fabric, cut on the bias, used to create a finished edge around curves.
- Elastic is available in various weights and is used for waistbands, underwear and swimwear.
- Ribbon, sequins and beads are applied to fabric surfaces for decorative effect.

Structural integrity   REVISED

Different methods are used to add structure to textile products to improve their aesthetics and function.

- **Boning** uses plastic or metal boning strips, which are sewn into reinforced seams to support fabric, preventing creasing and buckling. It is used in corsets, including medical corsets where it is used to protect people with injuries by restricting movement.
- **Layering** can be used to add structural support, insulation, comfort or body to a textile product.
- **Interfacing** is a non-woven bonded fabric that can be sewn or ironed on to the outer fabric or lining to provide light support, to stiffen fabrics and to improve shape. It is often used around necklines, armholes, waistbands, collars and cuffs.
- **Interlining** is a layer of fabric added between the fabric and the lining of a garment to add insulation as well as support. It is often used in suit jackets and winter coats.
- **Lining** is a lightweight, usually silky, fabric used inside a garment to improve comfort and to hide seams and other construction methods.
- **Underlining** is a layer of fabric used beneath sheer fabric to provide opacity or to add extra body to a garment.

Figure 19.12 Boning sewn into reinforced seams

Finishes and surface treatments

During the production of fabrics, finishes can be added to improve the aesthetics, comfort or function. These finishes are applied mechanically, chemically or biologically.

Dyeing

- Dyeing adds colour to fibres and fabrics using natural or synthetic dyes.
- Natural dyes work well on natural fibres and can be made from plants, minerals or insects.
- Synthetic dyes work on synthetic fibres, but can also give deeper or brighter colours when used with natural fibres.
- Fibres and fabrics can be dyed at polymer, fibre, yarn, piece or garment stage.

Clipping seams

- Clipping uses shears or specialist tools to cut into the seam allowance around necklines and armholes to achieve a neat finish and allow the fabric to lie flat.

Mechanical, chemical and biological finishes

Table 19.8 Mechanical, chemical and biological finishes

Finish	Description
Mechanical finishes	
Brushing	Wire brushes are passed over the surface of the fabric to raise the fibres and to produce a soft, fluffy surface. Often used on fleece and flannel fabrics.
Calendering	Fabric is pressed using heated rollers to give it a smoother, more lustrous surface.Often used on upholstery fabrics to give them a flat surface and sheen.
Chemical finishes	
Mercerising	Uses caustic soda to cause fibres to swell up, creating a more lustrous, stronger fabric. It improves the uptake of dye, giving a deeper and more even colour. Only works for cellulose fibres.
Crease resistance	A resin coating is applied that reduces absorbency and stiffens fibres so a fabric is easier to care for, reducing the need for ironing and allowing the fabric to dry more quickly.
Flame resistance	Applied to the surface of fabrics such as soft furnishings, children's sleepwear and bedding as a liquid coating; when dry this is durable and long lasting. Proban is an example.
Bleaching	Removes natural colour and prepares fabric for dyeing and printing to ensure an even, consistent colour.
Biological finishes	
Stonewashing	Used in the production of denim jeans. A popular distressed look is created by adding stones to industrial washing machines, along with the jeans.

Smart finishes

- **Thermochromic** inks change colour with changes in temperature. They can be printed on to fabrics and are used on baby clothes to detect changes in body temperature.
- **Photochromic** inks change colour with changes in natural light. They are printed on to the surface of fabrics and come in a range of bright colours. They are used on children's clothing and hotel sunbeds to alert people to sun exposure.
- **Microencapsulation** adds tiny bubbles or capsules containing chemicals such as insect-repellent, perfumes, moisturisers or antibacterial liquids that are added to the weave or fibre of fabric. They burst with friction to release the chemicals.

Thermochromic: Products that change colour with changes in temperature.

Photochromic: Products that change colour with changes in natural light.

Microencapsulation: The process of adding tiny bubbles or capsules that contain chemicals to the weave or fibre of fabric. These burst with friction, releasing the chemical inside.

Figure 19.13 **Microencapsulation – odour-controlling capsules added to trainers**

Surface decoration

Surface decoration techniques can be used to improve the aesthetics of a product, by adding colour, texture and pattern.

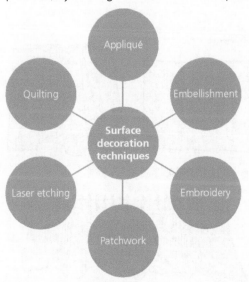

Figure 19.14 **Surface decoration techniques**

Processes used to make early iterative models

- Iterative models allow a designer to test, manipulate and adapt design ideas.
- A **toile** is a garment **prototype** made in cotton calico fabric (or other cheap fabric). It can be produced using basic pattern blocks or by following a simple commercial pattern.
- Toiles allow a textile designer to test construction methods that will be used on their final product.
- Different types of seams, shaping and decorative methods can be tested by creating small samples.
- Tests on materials to assess their abrasion resistance, stain resistance, absorbency, strength and insulation are carried out.
- Components are also tested and compared for their size, weight, function and aesthetic.

> **Prototype:** A test model on which the final product is based.

Figure 19.15 Toiles made from calico and newspaper

Manipulating and joining

Seams

- **Plain seams** are the most commonly used seam and are suitable for most fabric types.
- **French seams** are enclosed seams that hide any raw edges; they are used on sheer materials such as chiffon.
- **Flat felled seams** are very strong, enclosed seams used on denim and sportswear.

Seams are finished to neaten them and to prevent fraying by overlocking, pinking shears or a zig zag machine stitch.

Deforming and reforming

Shape or body can be added to fabrics to add interest and improve fit and function using the following methods:

- **Pleats:** the fabric is folded back on itself and sewn into place to add shape or body.
- **Gathers:** a long stitch length is sewn along the edge of the fabric in two rows. The thread ends are then pulled to create the gathers, which give fullness to a garment.
- **Darts:** folds are created in the fabric that taper to a point to shape and improve fit.

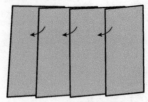

Figure 19.16 Pleats

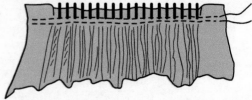

Figure 19.17 Gathers are an easy way to add shape to a product

Ensuring accuracy

Pattern marking and cutting

- Commercial pattern pieces and basic blocks are printed with a range of markings that must be followed accurately to ensure that the finished product is the correct size, shape and quality.
- Pattern markers can use tailor's chalk, vanishing markers, a tracing wheel and carbon transfer or tailor's tacks to transfer pattern markings on to fabric accurately.
- The paper or card template is placed on the surface of the fabric and cut around to produce fabric pieces that are of accurate size and shape.
- Commercial patterns contain templates that are printed on large sheets of pattern paper.
- Basic blocks are a basic template that can be adapted to produce a range of products.
- Commercial pattern pieces and basic blocks have an arrow to show the direction of the **grain line,** which tells the pattern cutter how the pieces should be placed on the fabric.
- The straight grain is usually used in garment production to provide the best drape; it runs vertically down the warp yarn in the fabric.
- The bias is at a 45-degree angle to the straight grain. A fabric cut on the bias is more flexible and can be sewn into curved shapes, but this method does create more wastage.

Fabric can be cut using fabric shears, a rotary cutter or laser cutting (usually used for more complicated patterns).

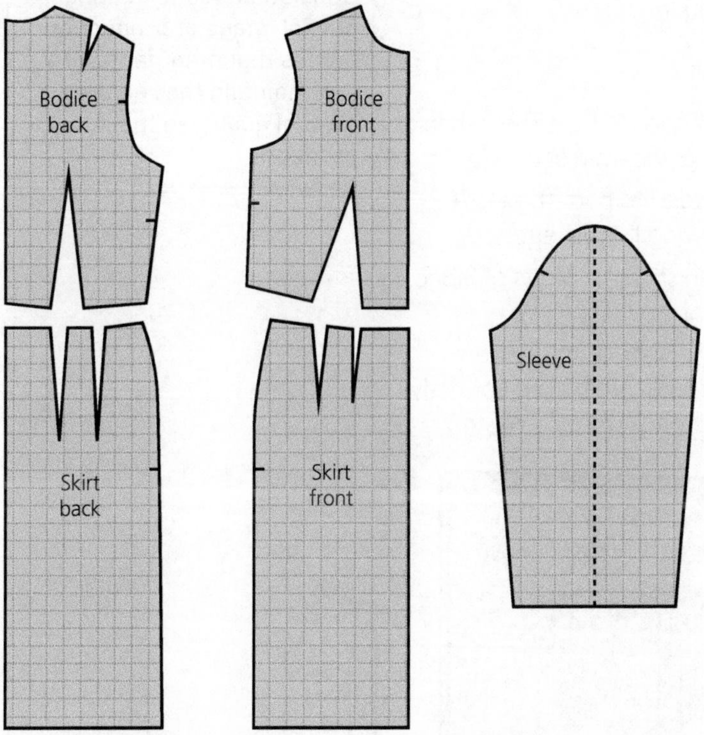

> **Grain line:** The grain line always follows the direction of the warp yarn, and shows where the pieces should be placed on the fabric prior to cutting.

Figure 19.18 **Basic blocks can be adapted to suit many patterns**

Quality control checks

- **Quality control** checks are carried out at critical control points to ensure consistency in the finished product. They identify:
  - faults with materials (for example snags, misprinted designs, holes or stains)
  - faults with or incorrect use of components such as buttons and zips
  - whether seams are within tolerance and that they are straight and have no holes
  - whether logos, pockets and sleeves have been placed correctly.
- NACERAP is an example of a quality control system.

> **Quality control:** Sample products are taken off the production line at critical control points to check for faults or to test performance.

Table 19.9 **The stages of NACERAP quality control checks**

	Example: Shirt
N (Name of fault)	Misaligned buttonhole
A (Appearance)	Misaligned buttonhole
C (Cause)	Error in measurement for button spacing on manufacturing specification
E (Effect)	Hem of shirt does not meet when buttoned up
R (Repair)	Unpick buttons and re-sew using new measurements
A (Action)	Make changes to button spacing measurement on manufacturing specification
P (Prevent)	Check all measurements on manufacturing specification for further errors

Quality assurance

Quality assurance means that consistent quality testing has been carried out at each stage of production and that manufacturers are able to maintain the required level of quality in their products.

> **Quality assurance:** Consistent quality testing at each stage of production means that manufacturers can maintain the required level of quality in their products.

Tolerances and minimising waste

- A **tolerance** is an allowable amount of variation of a specified size. These should be small (+/−1 cm) to ensure finished products are not too small or too big.
- In garment production, seam allowances are added to pattern pieces to allow enough fabric around the edge of the sewing line for errors.
- Wastage is created during pattern cutting, when small amounts of fabric are left around each pattern piece.
- Minimising this waste is the responsibility of lay planners, and pattern cutters have a responsibility to ensure pattern pieces are placed correctly on the fabric, leaving minimal space between them to reduce waste.

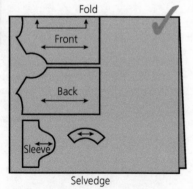
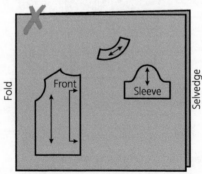
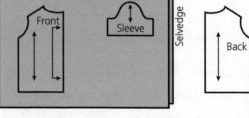

All grain lines are followed correctly.
Minimal wastage between pattern pieces.
Leftover fabric is a useable size and shape.
Fold marking has been followed correctly.

Large areas of unused fabric.
Not all pattern pieces fit.
Grain line has not been followed on all pieces.
Fold marking has been ignored.

Figure 19.19 **Pattern pieces must be placed correctly to minimise wastage**

Digital design tools

Rapid prototyping

- Rapid prototyping in textiles is still in the early stages of development but it is possible that 3D-printed clothing could be seen in the future.
- Rapid prototyping uses a laser cuter to cut layers of fabric, which are bonded together to create 3D forms.

Digital manufacture

- **Digital fabric printing** uses a large-scale inkjet printer and specialist dyes to transfer a digital image to the surface of the fabric.
- **Sublimation printing** uses heat and pressure to transfer dye from specialist printer paper on to the fabric.
- **Laser cutting** is computer controlled, allowing a 2D image to be drawn and cut quickly and accurately. Sometimes the surface of the fabric is etched with a laser cutter rather than cut entirely.

Interpretation of plans

Digital lay planning

- A digital **lay plan** uses computer software to plan where to place pattern pieces on fabric.
- Lay plans help a pattern cutter follow the grain line and minimise wastage.

Computer-aided design

- Image creation applications (such as Adobe Illustrator, CorelDraw and Digital Fashion Pro) allow textile designers to transfer hand-drawn sketches to a screen so that dimensions, shape and form can be refined.
- They give designers a 3D view of their sketch, which can be evaluated before prototyping.
- Designs for printing can be developed and transferred to CAM systems.

Computer-aided manufacturing

- CAM is widely used in textiles manufacture for printing, cutting and joining fabric; a wide range of semi- and fully automated machinery is available.
- CAM ensures consistent, accurate textile products can be produced quickly.

Figure 19.20 Laser cutting can be used to add intricate patterns to fabric

> **Lay plans:** Used by pattern cutters like a map to guide them when placing pattern pieces in the correct location and direction before cutting. Lay plans are carefully designed to minimise wastage.

Manufacturing at different scales of production and processes used

Manufacturing methods vary depending on the number of items being produced, timescale and budget.

Scales of production

Table 19.10 **Scales of production**

Production method	Description	Examples
One-off, bespoke production	• One-off, or bespoke, products are made by highly skilled workers – often an individual or a small team. • Products made in this way take a long time to produce as most of the work is done by hand. • The consumer usually has opportunities to attend fittings and make design decisions during the manufacturing process. • This type of product is often very high quality, and therefore expensive to purchase.	Wedding dresses
Batch production	• Batch-produced products are made by large teams of workers, working at various stages around the factory. • This type of production utilises a mix of semi-automated machinery and hand assembly. • Workers are specialised in one element of the construction process, such as collars or hems. Each employee works through a batch of partial products, which are then passed around the production line until they are complete. Batch-produced products are usually of mid- to low quality.	Seasonal clothing, (e.g. summer dresses and winter coats)
Mass production	• Mass production is the largest scale of production available. • This method is used for products that are in consistently high demand. • Many factories run 24 hours per day in order to maximise output and profit. • CAM is used widely in mass production as consistency and speed are so important. • Quality is controlled via computer so instances of faulty products are low and products are consistent.	Socks; plain T-shirts
Lean manufacturing	• A manufacturing method designed to minimise waste at each stage of production.	Lean manufacturing can be applied to any product by reducing set-up or change-over times during the manufacturing process, reassigning staff to other jobs during any downtime, or reducing the range of products available.
Just-in-time manufacturing	• A manufacturing method that can respond to changes in trends quickly by ordering materials and components to arrive at the factory 'just in time' for production. • This method minimises storage space, wasted materials and leftover stock, while increasing efficiency.	Often used by small businesses, for example those specialising in on-demand merchandise where products es, for example those specialising in T-shirts, hats and bags are only printed when the customer has placed the order.

Bespoke production: Manufacture of 'one-off' products that are designed and made for a specific client by an individual or small team of highly skilled workers. Bespoke products are high quality, can be complex and are expensive to make. Examples of bespoke products include wedding dresses, tailored suits and custom-fit car seat covers.

Batch production: Batch-produced products are manufactured by a large team of workers who each complete a specific stage of the production. These products are usually consistent in quality, available in a range of styles and sizes, and fall into the mid-to-low price range. Typical products include summer dresses, fashion T-shirts and branded school bags.

Mass production: Mass-produced products are manufactured mostly on automated machinery, operated by teams of workers. Products are manufactured very quickly, are consistent in quality and the range of available styles is minimal. Mass production costs are much lower than other production methods. Examples of mass–produced products include plain socks, plain T-shirts and plain baseball caps.

Lean manufacturing: A manufacturing method designed to minimise waste at each stage of production.

Just-in-time manufacturing: A manufacturing method that can quickly respond to changes in trends by ordering materials and components to arrive at the factory 'just in time' for production. This method minimises storage space, wasted materials and leftover stock, while increasing efficiency.

Figure 19.21 Wedding dresses are often bespoke

Manufacturing processes used for larger scales of production

Band saw cutting

- Band saws are used for large-scale pattern cutting.
- They can cut through up to 100 layers of fabric accurately and quickly.

Flatbed printing

- Flatbed screen printing uses a series of silk screens as wide as the fabric width, which are attached to the printer; each screen applies a different colour to the fabric.
- A conveyor belt moves fabric underneath the screens; when the belt stops, the screens drop down on to the fabric and an automated squeegee drags ink across the screen before the screen lifts and the fabric moves on.

Figure 19.22 Band saws can be used in industrial factories to cut through up to 100 pieces of fabric at a time

Rotary screen printing

- Rotary screen printing is also carried out on a conveyor belt production line.
- It uses printing cylinders instead of flat screens and therefore is quicker and takes up less space.
- The cylinder is filled with ink and a squeegee is placed inside.
- The cylinders spin as the fabric passes underneath, printing a continuous pattern on to the surface, with each printing a single colour and design.
- This printing method is expensive.

Industrial sewing machines and overlockers

- Industrial sewing machines and overlockers are heavy-duty machines with large motors that are designed to withstand constant use.
- They are fast and use large spools of strong thread to minimise snapping.
- They can sew through heavy or tough materials easily.

Automated presses

- Automated presses are used to remove creases and to finish completed garments before packaging.
- The machine has pressing plates, between which garments are placed. When they come together, they apply pressure, steam and heat to the garment.

Steam dollies

- Steam dollies are used to press garments that are unsuitable for automated presses because of their shape.
- A steam dolly is like a mannequin: the garment is placed on to it and it releases steam. It removes creases while maintaining the shape of the garment.

Cost and availability of materials and components

REVISED

The significance of cost

- It is important to consider the cost of materials and components when making design decisions: materials and components chosen must be within your (or your stakeholder's) budget and allow you to make a profit.
- You may need to adapt your design (for example, scale down, simplify or substitute) due to cost constraints.

Calculating quantities, costs and sizes of materials

- Measure the height of the lay plan (along the cross grain) and multiply by two. This is the minimum fabric width required.
- Measure the width of the lay plan (along the straight grain); this tells you how many metres of fabric you need to buy.
- In batch and mass production, fabric and components are ordered in bulk from wholesalers. The price drops as the number of units purchased increases.

> **Exam tip**
>
> Maths is important when working out quantities and tolerances. Make sure you can calculate measurements or take a calculator into the exam.

Now test yourself

TESTED

1 Describe one example of a plain weave fabric. [1 mark]
2 'Synthetic fibres are made from polymer chains' true or false? [1 mark]
3 Herringbone and houndstooth fabrics are examples of what type of weave? [1 mark]
4 Give two consequences of 'fast fashion'. [2 marks]
5 Give one benefit of adding interfacing to a garment. [1 mark]

20 Electronic and mechanical systems and control (design engineering)

Sources and origins

- Electronic and mechanical components are manufactured using a range of materials and processes.
- System components are manufactured all around the world on a very large scale by specialist manufacturers.
- A range of raw materials and chemicals are required, which are difficult and dangerous to extract and process.
- As these system components become more complex, the number of companies able to manufacture them decreases.
- 'Chip famine' occurs when there is a shortage of a certain type of microchip due to there being a high demand that cannot be met.

Lifecycle

- Most electronic components are extremely reliable in use and rarely fail, so they have a long lifespan if used correctly.
- If a component is subjected to higher than its intended voltage or current, this can cause premature failure.
- Rechargeable batteries are the most common component with a limited lifespan. Over time they will gradually fail to hold a charge.
- Some components, such as normal batteries, are designed to be disposable but can be recycled as they are made from common metals.
- Mechanical components have a much shorter lifespan because friction causes them to wear away gradually.
- Lubrication helps to reduce friction and slows down the process, making them last longer.

Recycling, reuse and disposal

- Electronic and mechanical products can contain hundreds of different components, which are difficult to separate for recycling.
- Many of the components contain hazardous chemicals that are harmful to the environment if they are put in landfill.
- The Waste Electrical and Electronic Equipment (WEEE) directive requires manufacturers to 'take back' unwanted electrical products and old batteries and ensure they are properly recycled.

Standard units of measurement

- Mechanisms control or change motion.
- Mechanisms have an input force and an output force.
- Mechanisms can also increase and decrease output force or distance moved.
- **Mechanical advantage** (MA) is when a mechanism is used to either:
  - ○ increase the output force but reduce the distance it is moved, or
  - ○ increase the distance moved but reduce the force being exerted.

> **Mechanical advantage:** Using a mechanism to increase output force.

input force × distance moved by input = output force × distance moved by output

The mechanical advantage is expressed as a ratio:

$$MA = \frac{\text{output force}}{\text{input force}}$$

- Levers are one of the simplest mechanisms and use mechanical advantage to increase forces.
- Mechanical advantage for levers is calculated by dividing the load by the effort:

$$MA = \frac{\text{load}}{\text{effort}}$$

or the length of the input arm by the length of the output arm:

$$MA = \frac{\text{input arm length}}{\text{output arm length}}$$

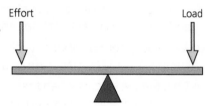

Figure 20.1 A lever with a pivot in the middle

- If a lever pivots in the middle, the input and output force and the distance travelled will be exactly the same, so there will be no mechanical advantage.
- If the fulcrum is moved nearer to the output end, then less force is required to lift the object, but it will not move as far.
- The position of the fulcrum also affects the type of lever.
- There are three classes of lever depending on the position of the fulcrum in relation to the load and effort.

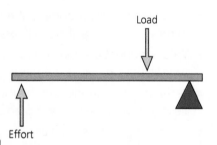

Figure 20.2 A lever with the fulcrum near output end

Table 20.1 The three classes of lever

First class lever	Second class lever	Third class lever
Fulcrum between the load and effort, e.g. bicycle brake lever, pliers, scissors	Load between fulcrum and effort, e.g. wheelbarrow, nutcracker	Effort between load and fulcrum, e.g. tweezers, staple remover
Scissors — Fulcrum, Effort, Load	**Nutcracker** — Fulcrum, Load, Effort	**Tweezers** — Fulcrum, Effort, Load, Effort

Electrical systems

- Electricity has a voltage and a current:
  - Voltage is the amount of electrical 'pressure' causing a current to flow, measured in volts (V).
  - Current is the amount of actual electricity flowing, measured in amps (A).
- Electricity flows through some materials more easily than others. The resistance of a material is how hard it is for electricity to flow through it.
- Resistance is measured in ohms (Ω).
- All electronic components have a different rating, which shows the maximum amount of voltage and current they can take without being damaged.
- All components also have a certain amount of resistance to the electricity flowing through them.
- Volts, amps and ohms are all related by Ohm's law:

$$\text{current (A)} = \frac{\text{voltage (V)}}{\text{resistance (Ω)}}$$

The formula can be rearranged to calculate voltage, current or resistance:

$$\text{voltage (V)} = \text{current (A)} \times \text{resistance (Ω)}$$

$$\text{resistance (Ω)} = \frac{\text{voltage (V)}}{\text{current (A)}}$$

For example, a light bulb could be rated at 12 volts, 0.5 amps, therefore:

$$\text{resistance} = \frac{12}{0.5} = 24\ Ω$$

Mechanical systems

- Rotational mechanical systems use rotational force or turning effect, which is called **torque**.
- They can either:
  - reduce rotational speed but increase torque, or
  - increase rotational speed but reduce torque.

Mechanical advantage is expressed as a ratio

$$\text{MA} = \frac{\text{output torque}}{\text{input torque}} \quad \text{or} \quad \frac{\text{input rotational speed}}{\text{output rotational speed}}$$

- **Gear trains** and pulleys are used to create this rotational mechanical advantage in mechanical systems.
- Mechanical advantage is also known as the **gear ratio**, which is calculated by:

$$\text{MA} = \frac{\text{number of teeth on driven gear}}{\text{number of teeth on driver gear}}$$

Typical mistake

Many electronic components are highly sensitive and built to operate between certain voltage and current levels. Don't be tempted to increase voltages without first testing levels with a multimeter.

Torque: Rotational force.

Gear train: Two or more gears meshed together.

Gear ratio: The diameter or number of teeth on the driven divided by the driver.

Standard components

- Gears are rotating mechanical parts with 'teeth' that mesh with other toothed parts to transmit movement and torque.
- Two or more gears meshed together are called a gear train.

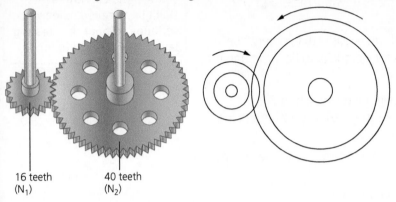

16 teeth
(N_1)

40 teeth
(N_2)

Figure 20.3 A simple gear train

- The input shaft has 16 teeth and the output shaft has 40 teeth, therefore the gear ratio, or mechanical advantage, would be:

$$\frac{\text{number of teeth on driven gear}}{\text{number of teeth on driver gear}} = \frac{40}{16} = 2.5, \text{ or ratio of 2.5:1}$$

- Because the amount of torque has been increased by 2.5 times, the speed of the output shaft would be reduced by the same amount so would turn 2.5 times slower than the input shaft.

Idler gears

- **Idler gears** are inserted into gear trains to reverse the direction of the output shaft.
- They are often used in gear trains to keep the rotation of the input shaft the same as the output shaft.
- Idler gears can be any size but do not affect the gear ratio of the gear train, whatever size they are.

> **Idler gear:** An extra gear used to keep the driver and driven gear turning in the same direction.

Bevel gears

- Bevel gears have teeth cut on a 45-degree angle.
- Bevel gears transmit rotary motion through 90 degrees and can also have different numbers of teeth to alter gear ratios.

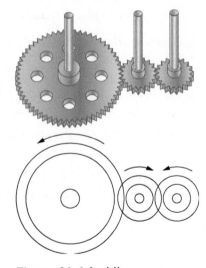

Figure 20.4 An idler gear

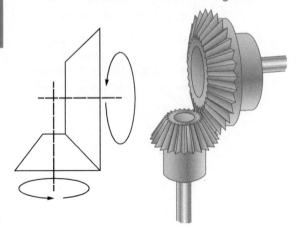

Figure 20.5 A bevel gear

20 Electronic and mechanical systems and control (design engineering)

Worm drives

- A worm drive gear has one tooth that is shaped like a screw or bolt thread.
- For each rotation of the shaft, the worm gear will move a meshed gear by only one tooth.
- Worm gears are used for transferring motion through 90 degrees.
- They are extremely useful for increasing torque or lowering speed by large amounts.
- It is easy to calculate the gear ratio as the driver gear has only one tooth.
- Worm drives only work one way, that is, the worm drive must be the driven gear.

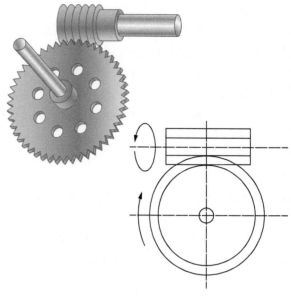

Figure 20.6 **A worm drive**

Rack and pinions

- Rack and pinions change rotary motion into linear motion or vice versa.
- The pinion has a normal spur gear with teeth around the outer edge, while the rack is a flat bar with teeth in a line.
- When the pinion turns, the teeth mesh and the rack moves. In the same way, moving the rack rotates the gear on the pinion.

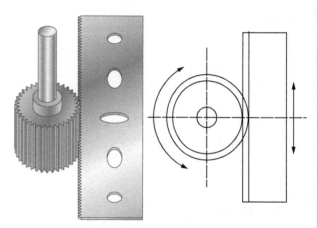

Figure 20.7 **A rack and pinion drive**

Chain and sprockets

- A chain is a method of transferring motion between two gears without the use of a gear train.
- The chain has links that mesh with the teeth on the sprockets and 'pull' the chain along.
- The driven sprocket, which is wrapped around the chain, is then rotated by the movement of the chain.
- The sizes of the two sprockets will be the gear ratio.

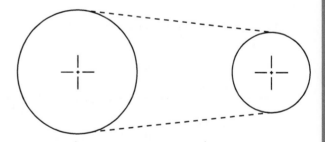

Figure 20.8 **Chain and sprocket**

Pulley and belt drives

Pulleys and belts work in a similar way to a chain and sprocket mechanism.

The gear ratio is calculated in a similar way, except the sizes of the pulleys are used instead of the number of teeth.

For example:

$$\text{gear ratio} = \frac{\text{diameter of driven pulley}}{\text{diameter of driver pulley}}$$

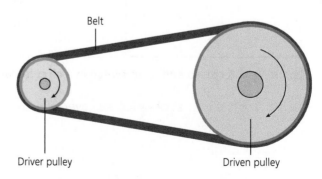

Belt

Driver pulley

Driven pulley

Figure 20.9 **A pulley system**

Pulleys and belts are quieter than chain and sprockets but can slip as the belt stretches in use.

To prevent this, many pulleys use:

● V-shaped belts and pulleys, so the belt can grip the side wall of the pulley
● toothed belts and pulleys similar to chains
● tensioner pulleys, which are extra pulleys that press on the belt to keep it the same tension.

Cams

● Cams are used to convert rotary motion to reciprocating motion.
● Cams do not alter gear ratios or mechanical advantage.
● The distance a cam rises is called the stroke.
● Cams are often used in engines to open and close valves.

Electronic components

Sensors

● There are two main types of sensor:
  ○ Digital sensors that detect on/off or yes/no situations, for example if the button is pressed or not.
  ○ Analogue sensors that measure the amount of something, for example the temperature in a room.

Switch sensors

● Switch sensors are digital and can detect whether different things are on or off.
● A closed switch is 'on' and an open switch is 'off'.
● Open means current cannot flow.

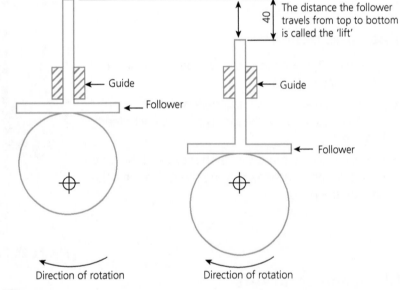

Figure 20.10 **Diagram of a cam**

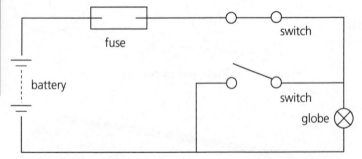

Figure 20.11 **A closed and open switch in a circuit diagram**

The many different types of switch are covered in Chapter 13.

Light sensors

Light-dependent resistors (LDRs) have a high resistance to electric current when there is no light. As the light level increases, the resistance decreases and more electric current is allowed to pass through.

Infrared sensors

There are different types of infrared (IR) sensors available:

- IR sensors that detect the presence of warm objects, for example hand dryers.
- Passive IR sensors that detect moving warm objects, for example movement sensors in a house alarm.
- IR distance sensors that measure the distance to objects, for example radar speed guns.
- IR receivers that pick up data signals from an IR transmitter, for example a TV remote.

The different sensors are covered in Chapter 13.

Output devices

Common output devices include:

- light-emitting diodes (LEDs)
- light bulbs
- speakers and buzzers
- motors.

These are all covered in Chapter 13.

Microcontrollers

Microcontrollers are used with electronic components to create electronic circuits and systems.

A microcontroller is a subsystem that is integrated into the electronic circuit.

Microcontrollers can be connected together to create more complex systems. This is called interfacing.

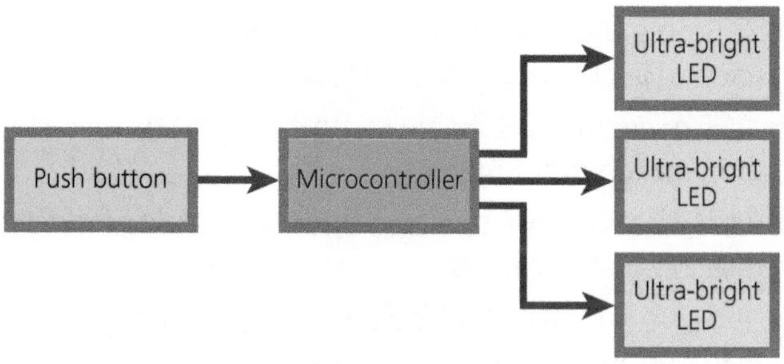

For example, a bicycle light has three LEDs that are turned on/off with a push button switch. By using a **microprocessor** that is programmed to control the lights in the circuit, lights can be made to work in different ways:

- first push of the button turns on all three LEDs
- second push turns on each one in turn for half a second at a time
- third push flashes all three LEDs
- fourth push turns all three LEDs off.

> **Exam tip**
>
> LEDs are a type of diode and only work when wired up one way.
>
> Reed switches use a magnet to open and close the contacts.

Figure 20.12 **A microprocessor in a system diagram**

> **Microprocessor:** Electronic system used to control output devices.

The cycle then repeats.

- Microcontroller programs are step-by-step instructions carried out in a set sequence.
- The sequence is carried out in a fraction of a second when the program runs.
- Flowcharts are often used to plan the sequence of steps the microcontroller will follow.
- Flowcharts help the designer to think like a microcontroller operates.
- Flowcharts use five different symbols to represent different operations.

Types of microcontroller

- The type of microcontroller used in a system will depend on:
  o the technical requirements of the system, such as the number of inputs/outputs needed
  o how easy or difficult the microcontroller is to program
  o the choice of power supply
  o the range of compatible components, such as light sensors, available.
- Many microcontrollers use complex language that requires specialist programmers.
- There are many simpler microcontrollers available for users with little knowledge and experience, which are ideal for use in GCSE work.
- New and more advanced microcontrollers that can be programmed by young children are also being developed and used in many primary schools to teach the basic functions.

Some of the most common microcontrollers used in schools are listed below:

- Crumble uses a language that is easy to learn and connects with crocodile clips or is soldered together.
- BBC micro:bit uses simple commands that teach children how to code and can control systems.
- PICAXE and GENIE can be programmed using flowcharts or BASIC and use integrated circuits.
- Arduino is more advanced and difficult to program but is powerful and can perform many complex instructions.

Symbol	Name
	Start/end
→	Arrows
	Input/output
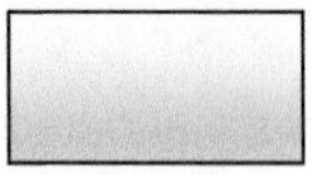	Process
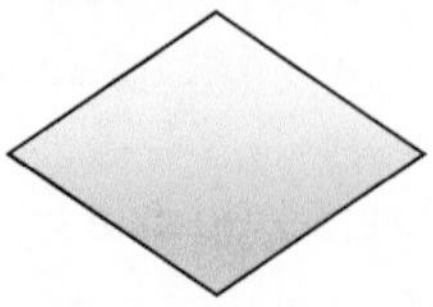	Decision

Figure 20.13 Symbols used in a flowchart

> **Typical mistake**
>
> Don't create complex programs for microcontrollers all in one go. Develop the program in stages and test each one before moving on to the next. This will make fault finding much easier.

Structural integrity

REVISED

- The structural support for a mechanical device or mechanism is often called a chassis.
- The chassis must be rigid and able to withstand the loads and forces placed on it by the mechanical movement.
- If a chassis is not sufficiently rigid, the mechanism will fail and may be damaged as a result.
- Many chassis use a hollow or box section to increase rigidity without adding excessive weight.
- Ribs can also be moulded into the chassis shape to increase rigidity while only adding minimal weight.

Triangulation

- **Triangulation** is used to add strength to frameworks that house and support mechanisms.
- Triangulation uses the natural rigidity of triangles to stop frameworks from collapsing by:
  - adding diagonal cross members across rectangular frames to divide them into two triangles
  - adding triangular gusset plates to corner joints.

> **Triangulation:** Adding strength to frames by adding triangle shapes.

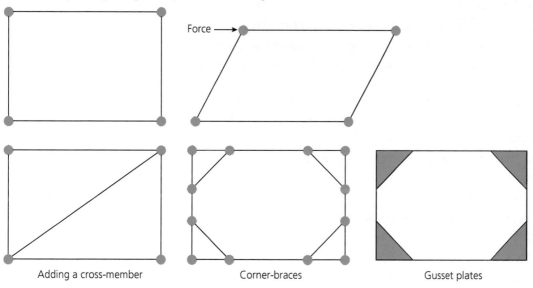

Adding a cross-member   Corner-braces   Gusset plates

Figure 20.14 Triangulation

Finishes and surface treatments

- Mechanical and electrical systems often work or are used outside, where they are exposed to the elements.
- If left exposed, moisture will quickly cause corrosion and damage to the system.
- Total waterproofing is possible, but sealing of casings must allow easy access for servicing and repair.
- Making splash-proof casings is much easier and usually sufficient for the majority of applications, using a rubber seal around the casing.
- Cable glands allow holes in cases where cables enter or exit to stay waterproof.
- Pre-lubed and sealed bearings are available that prevent dirt and moisture causing damage to mechanisms.
- Correct instrumentation labelling on electronic and mechanical systems is extremely important.
- Transfer labels, engraving and stick-on vinyl can be used to improve surface finishes and provide clear instructions for users.

Processes used to make early iterative models

- Models are often made to test the feasibility or effectiveness of a system or part of a system:
  - Physical models are made with real materials and components on a smaller scale to test if mechanisms will work as intended.
  - Computer-aided models such as 3D CAD programs and simulations are used in engineering.
  - Software models test flowchart programs for microcontrollers.
  - Mathematical models can test hypotheses and make predictions.
- Mechanical models that allow different gear ratios and combinations to be tested can be assembled using pre-manufactured components, for example Meccano and Lego Technic.
- Other parts, such as levers, linkages and cams, can be designed on CAD and cut out on the laser cutter or other CAM machine in card, Styrofoam, thin plastic or MDF.
- Elastic bands and split pins can be used to make models involving pulleys and pivots.
- Electronic models are used to:
  - check and test interfacing between the input/output devices and the microcontroller
  - check that the input and output devices function as intended
  - check the correct functioning of the flowchart program.
- Prototyping boards or 'breadboards' allows circuits to be built quickly, tested and modified without soldering.
- CAD software can be used to model circuits using a computer simulation without actually buying or joining any components, and allows components to be changed quickly and easily.
- Allowing real users to test microcontroller programs is essential to see how user friendly they are and whether they can be used by people who are not programmers or experts.

Manipulating and joining

- Electronic products have the circuits mounted on circuit boards.
- The circuit board holds the components in place and connects them together.
- Components are passed through holes in the board and soldered in place.
- Circuit boards are usually designed to be as small as possible.

Stripboard

- Stripboard has parallel rows of copper strips with holes at regular intervals.
- By soldering two components into the same strip, they are both connected by the copper.

Printed circuit boards

- Printed circuit boards (PCBs) are usually made from glass-reinforced plastic (GRP) and have a layer of copper on one side.
- Holes can be drilled anywhere on the PCB and the copper removed to create tracks that connect the necessary components together.
- The copper can be removed in two ways:
  o acid etching
  o isolation routing.

Acid etching

- Acid etching uses an acid (ferric chloride) to dissolve the unwanted copper areas on the board.
- The copper areas that are needed are covered by either:
  o drawing on to the copper using a special pen, or
  o painting over the copper with a suitable paint, such as enamel.

The PCB is then placed into a tank of acid which dissolves all of the copper that has not been covered.

Photo etching

- Photo etching works in a similar way to acid etching but the PCB has a thin layer of photosensitive film over the copper.
- The design of the tracks is drawn on the computer and printed in black ink on to a transparent film that is then placed over the copper.
- Ultraviolet (UV) light is shone through the film, hitting the areas not covered by the black ink.
- The PCB is then placed into the acid where the copper is dissolved from the areas that the UV light affected, leaving the other areas intact.

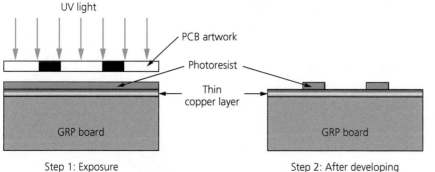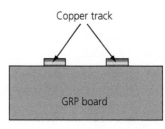

| Step 1: Exposure | Step 2: After developing | Step 2: After etching |

Figure 20.15 Photo etching

Isolation routing (PCB engraving)

- Isolation routing uses a CNC engraving machine to remove the copper areas from the PCB.
- The PCB is clamped firmly on to a bed and the CNC router follows a program generated by PCB design software.
- The router also drills holes for components in the necessary places.

 "20 Electronic and mechanical systems and control (design engineering)"

20 Electronic and mechanical systems and control (design engineering)

Soldering

- Soldering uses melted metal to securely join and hold wires and components in place.
- Solder is an alloy of tin and copper with a relatively low melting point compared to other metals.
- Solder is in the form of a thin wire that also contains flux (a chemical cleaning agent to help make a good joint).
- Joints between components are heated with a soldering iron and solder is added, which melts and flows around the joint.
- Once the heat is removed, the solder cools quickly and hardens, forming a solid joint.
- Soldering irons come in a range of sizes and heat ratings, depending on the size of the components being used.
- Poor soldering will result in 'dry' joints that will fracture or cause poor connections during testing or later in use.

Mechanical joining systems

- Mechanical systems require gears, pulleys and other parts to be joined to shafts, axles and other parts.
- In industry and in large machinery where high torque and speeds are used, there are specialist methods of joining gears to axles and so on.
- For project work and iterative modelling at GCSE level, there are some simple ways of joining common mechanical components.

Attaching a wheel or gear to a shaft

- For low torque systems, a 'push fit' can be used; the gear has a hole fractionally smaller than the shaft and can be pushed on, achieving a tight fit.
- Splined shafts are similar to push fit but the splines grip the gear much more securely so it can be used in much higher torque applications.
- Grub screw fixings use a small headless screw that tightens against a flat section on the shaft.
- Grub screw couplings can be used to join shafts together in a straight line or at an angle.
- Universal couplings also use grub screw fixings and allow the transfer of movement through other angles.

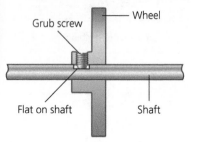

Figure 20.16 **Grub screw fixture for a gear**

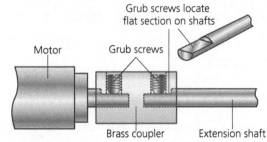

Figure 20.17 **Grub screw fixture for a coupling**

Other standard fasteners and fixings

There are many other standard fixings available such as nuts, bolts, washers, screws and so on.

Moulding

For information on moulding of metal (casting), see Chapter 17. For details of injection moulding and blow moulding, see Chapter 18.

Now test yourself answers and quick quizzes at **www.hoddereducation.co.uk/myrevisionnotes**

Accurate marking out methods

REVISED

For information about marking out see Chapters 16 and 17.

Digital design tools

REVISED

- Rapid prototyping allows parts to be designed on CAD and produced on CAM machinery in a matter of minutes or hours, depending on their complexity.
- Rapid prototyping is much quicker than conventional machining of parts and requires less skill.
- Complex parts that would be extremely difficult to machine by hand can be produced.
- Laser cutters can be useful for cutting parts from thin sheet materials, such as chassis parts and levers.
- Plasma cutters, CNC routers and milling machines can be used to make many different parts that have been designed on CAD.
- Autorouting software is used in PCB manufacture to design track layouts that optimise space on the board to make them as small as possible.
- Modern devices such as mobile phones would be much larger without autorouting software.

Manufacturing at different scales of production and processes used

REVISED

One-off, bespoke production

- Where a specialist mechanical part is required, this will be made as a one-off or 'bespoke' part.
- Bespoke production requires a high degree of manufacturing skill.
- Rapid prototyping is often used to make bespoke components.
- Microcontrollers are often programmed as one-offs for specific clients or for use in testing.
- One-off PCBs that are designed and assembled by hand are used in prototyping.

Batch production

- Batch production is used to manufacture components and assemblies for a small number of products.
- Batch manufacturing of parts for a simple mechanical product would usually take place over a few days, for example:
  - all chassis parts are manufactured
  - chassis are assembled
  - final components are added
  - product testing
  - packaging and dispatch.
- In batch production, every member of staff must be skilled in all processes.
- In cases where the complexity of production is extreme, it may be subcontracted to specialists.
- PCB production is often done in batches.

20 Electronic and mechanical systems and control (design engineering)

Mass production

- Components such as screws, bolts, nuts, connectors and batteries are mass produced in large numbers by specialist manufacturers.
- Mass production of these parts requires specialist equipment and tools capable of producing large volumes and running continuously.
- Mass production requires no staff with specialist manufacturing skills, but 24-hour monitoring by employees working shifts is required.

Lean/just-in-time production

- Lean manufacturing is a system of reducing waste at every stage of production.
- Reducing waste materials, time, space and resources are the main objectives.
- Just-in-time (JIT) production is one example of lean manufacturing.
- Manufacturers order materials and components so that they arrive just before they are needed.
- Storage space and facilities are saved, which in turn reduces manufacturing costs.
- The system depends on reliable suppliers that will deliver materials on the correct delivery dates.
- It only takes one supplier to miss a delivery to hold up the whole production process.
- Products are shipped as soon as manufacture is completed, so storage space is not required.

PCB manufacture

- Multi-layered and double-sided PCBs have components on both sides of the board and multiple boards on top of each other to save space.
- Extremely complex circuits can be constructed in very small spaces.
- Surface mount technology (SMT) is a method of PCB manufacture used for large-scale production.
- Tiny components, often less than 1 mm in size, are placed on to the board using pick and place machines.
- Pick-and-place machines accurately position and place components on the board at a rate of four or five per second.
- The board is coated in sticky solder paste that is then heated up to melt the solder and hold each component firmly in place.
- Optical recognition technology uses a computer to visually inspect and check that each component is correctly placed after manufacture.

Now test yourself

TESTED

1 Name one example of:
   a) a first class lever
   b) a second class lever
   c) a third class lever.   [3 marks]
2 Calculate the mechanical advantage of the following:
   a) Driver gear = 25 teeth; driven gear = 75 teeth.
   b) Driver pulley = 60 mm diameter; driven pulley = 12 mm diameter.   [2 marks]
3 Name one advantage of a chain compared to a pulley.   [1 mark]
4 Name two analogue sensors.   [2 marks]
5 Explain the difference between a speaker and buzzer.   [3 marks]

Now test yourself answers and quick quizzes at **www.hoddereducation.co.uk/myrevisionnotes**

Success in the examination

When will the exam be completed?

You will sit the paper at the end of the GCSE course in May or June of the second year of the course, which is usually in Year 11.

How long will I have to complete the exam?

This examination will be two hours long. The paper is divided into Sections A and B, and while both sections are available for the full two hours, you should use more time to answer the in-depth questions in Section B of the paper.

What type of questions will appear in the exam paper?

The exam paper will be split into two sections.

Section A will include three sets of questions that will test your 'core' knowledge. The questions will include

- a product analysis question looking at a multi-material product
- questions that test your mathematical skills
- a question that requires technical understanding.

There will be a mixture of different levels of questions and you will need to demonstrate a comprehensive and thorough knowledge of the topic. One question will be an extended response question, which you will be able to identify as it will have an asterisk (*) next to it. This question will look for an in-depth understanding of the topic, and you will need to write a detailed and logical answer that is supported by examples.

Section B will test both your core and in-depth knowledge. As in Section A, there will be a mixture of different levels of questions. These will expect you to demonstrate a comprehensive and thorough knowledge of the topic. There is one extended response question that will have an asterisk (*) next to it. You will also be given an insert document that will give details of products and a situation relating to the six categories of learning:

- papers and boards
- natural and manufactured timber
- ferrous and non-ferrous metals
- thermo and thermosetting polymers
- natural, synthetic, blended and mixed fibres, and woven, non-woven and knitted textiles
- design engineering.

One question will require you to choose a product in order to demonstrate your in-depth understanding in relation to any one of the main categories of materials and/or systems, including the sources, development and manufacture of prototypes and products using the chosen categories of materials and/or systems.

Tips on preparing for the exam

- If you did not understand a topic when it was covered in class, you are unlikely to understand it when revising. Make sure you ask at the end of a lesson if you're unsure about any of the material covered.
- Being absent from school can leave a big hole in your knowledge. Make sure that you catch up on any missed work.
- Don't leave revision to the end of the course. Test yourself at the end of each topic.
- Use past papers, online materials and revision guides to help you practise exam-type questions.
- Plan your revision time in the weeks leading up to the exam.
- Make revision cards to help you compartmentalise your understanding.
- Work with other students to test each other.

Approaching the paper

- Make sure you know the date, time and location of your exam.
- Get a good night's sleep. Make sure you have eaten and that you are hydrated.
- Arrive early and make sure you have all your equipment with you.
- Read the instructions on the front cover of the question paper. This will tell you what you have to do.
- Read each question carefully at least twice. This will help you to understand exactly what information you need to give.
- Each question will tell you how many marks are available for this question. Use this to gauge how much detail you need to put into your answer.
- The question will tell you how long you should take to answer the question. Use this to help you pace yourself through the exam.
- If you finish early, go back and reread the questions and your answers. You will usually find that you have remembered more detail. You may also be able to spot any mistakes that you may have made.

Exam practice questions

Paper and boards

1) The figure below shows a takeaway pizza box. Features of the box are listed below.

a) Complete the description of each feature. The first one has been done for you. [4]

Feature	Description
Takeaway pizza box has a flat top and bottom	So the boxes can be easily stacked on top of each other.
Square-shaped package	
Corrugated cardboard material	
Logo and instructions	
One piece net (development)	

b) Most papers and boards can be recycled after use. Give one reason why a takeaway pizza box may **not** be suitable for recycling after use. [1]

c) Mechanical pulp is made up from **two** ingredients: virgin pulp and recycled pulp. Virgin pulp accounts for 45 per cent of mechanical pulp. What percentage is recycled pulp? [1]

d) Use sketches and notes to show the four stages in the mechanical pulping process. [6]

Timbers

The figure below shows a storage unit made from oak.

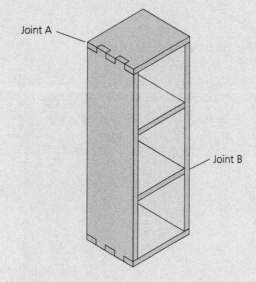

Joint A

Joint B

1 a) Name the corner joint (A) used on the storage unit.   [1]
  b) Name the joint (B) used for the shelves.   [1]
  c) Name one other suitable joint that could be used for the shelves.   [1]
2 Describe how to make the channels on the shelf joints, including the tools and equipment used.   [3]
3 State a suitable 'natural' finish for the storage unit.   [1]

Metals

The figure below shows a coat hook made from mild steel.

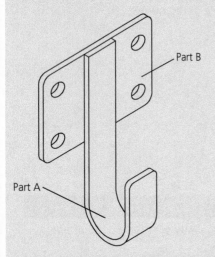

Part A is made from a length of flat bar.
The sectional view (profile) of the bar is shown in the table below.

1 Complete the table by drawing and naming three other stock profiles of mild steel. [3]

Flat bar			

2 State one suitable tool you could use to:
  a) Cut the mild steel bar to length.   [1]
  b) Round off the corners.   [1]

Polymers

The figures below show the prototype for a fruit juice bottle

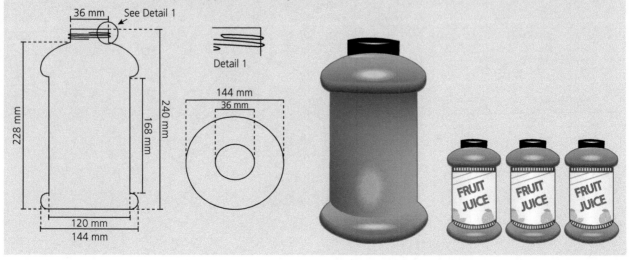

Detail 1

The three parts of the product are:
- a blow-moulded bottle
- an injection-moulded cap
- a printed label.

1 The blow-moulded bottle is made from polyethylene terephthalate (PET).
Give **two** reasons why PET is suitable for this product. [2]

2 Use sketches and notes to describe the process of blow moulding. [6]

3 Name a suitable polymer for the injection-moulded cap. [1]

4 Explain why injection moulding costs per unit decrease for high-volume production runs. [2]

5 The printed label is made from paper. It goes all the way round the part of the bottle that is 1 20 mm in diameter and overlaps by 10 mm.

a) State one reason why paper has been chosen for the label. [1]

b) State one reason why the printed label overlaps. [1]

c) Calculate the length of the label including the 10 mm overlap. [3]

6 'Labels must inform and protect the consumer.' Discuss this statement in relation to the fruit juice bottle label. [6]

Textiles

1 a) Give the names of two natural fibres. [2]

b) Give the names of two synthetic fibres. [2]

c) The figure below shows a hat made from a blended fibre. Explain the term 'blended fibre', giving an example. [2]

d) Name the fabric that is produced by applying heat, moisture and friction to fibres. [1]

e) Complete the table below to show the descriptions for the properties of fabric. The first one has been done for you. [5]

Property	Description
Insulating	Keeps us warm
Absorbent	
Durable	
Drape	
Elasticity	
Handle	

Design engineering (electronic systems)

The figure below shows a security light.

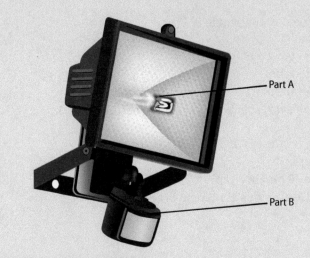

Part A

Part B

1 Name parts A and B of the security light.

Part A _____ [1]

Part B _____ [1]

2 Complete the following flowchart for the security light. [4]

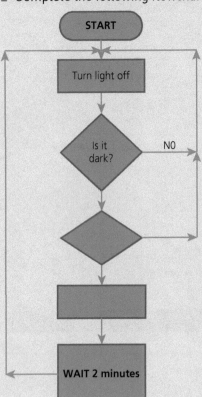

Sample examination questions

In this section you will see some examples of the types of questions you may get in your written examination. They show some candidates' responses and there is an explanation of why the marks were awarded.

Section A

Example

1

a) Complete the table below to show the category of material used to make parts of the products listed. [6]

Product	Material used
Handle of the trowel	
Watering can	
Blade for the shears	

b) Explain how the plant pot could be made more sustainable by applying the three Rs shown below. [6]

Reduce	
Recycle	
Reuse	

c) The seed packet is made from foil-backed paper. State **three** qualities of foil-backed paper that make it suitable for a seed packet. [3]

_____

_____

_____

Candidate response

1 a)

Product	Material used
Handle of the trowel	A hardwood
Watering can	A thermo polymer
Blade for the shears	A ferrous metal

b)

Reduce	The walls of the plant pot could be made thinner so that less material is used, which is good for the environment.
Recycle	Once the plant pot has reached the end of its useful life, it could be taken to a recycling centre so that the material can be thrown away nicely.
Reuse	Once the plant pot has become cracked or split, it could be cut down and used as a tray.

c) 1 Smooth paper surface for printing information.
   2 The foil makes the package waterproof so the seeds will not get damaged.
   3 Foil-backed paper is easy to cut, fold and join together.

a) The specific category of material is required rather than a material.
One mark for hard [1] and one mark for wood [1].
One mark for polymer [1] and one mark for thermo [1].
One mark for metal [1] and one mark for ferrous [1].

b) One mark for stating a change/action to the product and one for linking the change to sustainability.
Award of marks shown by ✔ and ✘.
Reduce: The walls of the plant pot could be made thinner ✔ so that less material is used, ✔ which is good for the environment.
Recycle: Once the plant pot has reached the end of its useful life, it could be taken to a recycling centre ✔ so that the material can be thrown away nicely. ✘
Reuse: Once the plant pot has become cracked or split, it could be cut down ✔ and used as a tray. ✔

c) One mark for each correct answer. All three answers are correct.

Section B

Example 1: Paper and boards

1* A designer is making a model of a new shopping centre.
The model will be made from foam board.
Discuss the ecological issues that the designer must consider. [8]

Candidate response

Foam board is made of a layer of polystyrene foam sandwiched between two layers of card. It cannot be recycled together and would need to be separated. This would be very difficult to do and there would also be an amount of glue (to hold the sandwiched layers together) that would also need to be removed. Foam board causes damage to the environment in both the manufacture and disposal of the material. The core layer of polystyrene causes ecological damage as the extraction of the raw material (oil) to manufacture the polystyrene is unsustainable. At the end of its lifecycle, foam board is usually put into landfill because of the difficulty in recycling. Landfill increases damage to our environment by increasing carbon emissions and leaking pollutants into the earth, sea and air. Laminated papers and boards that contain PVC can take up to 500 years to decompose and they can release toxic chemicals during their decomposing. Foam board usually has bright white sandwiched layers, which would indicate that the pulp had been bleached to obtain a very white finish. Bleaching can also be harmful to the environment in the manufacture, use and disposal of the chemical. However, foam board is a good material to use for model making. It is easy to cut, shape and join. It is strong yet lightweight; it is a smooth, sleek surface and printed finishes can be applied to it. When making a model of a new shopping centre, a designer may choose to use foam board for its physical properties, but they should be mindful of the ecological implications. As the model is likely to be a one-off, the benefits may outweigh the negatives.

Assessment comment

This is an example of a very good response that would achieve full marks. The candidate has shown that they understand what an ecological issue is and that they have knowledge of the material, foamboard, and its associated advantages and disadvantages. The candidate can make clear links between the material and its impact on the environment, both during production, use and disposal and link this to the designer. They have used technical terms (extraction of raw materials, pollutants, decomposing, bleaching, pulp), have shown specialist knowledge and the response is constructed in logical sentences to demonstrate the issues that the designer would face when selecting the material.

Mark scheme

Level 3 (6–8 marks)	The candidate will show good knowledge and understanding of the ecological issues that arise with the use of foamboard. There should be good consideration of the environmental connection to these issues in relation to the life cycle of a product or environmental impact. The candidate will be able to undertake a thorough evaluation of the issues, identifying positive and negative implications. There is a well-developed line of reasoning which is clear and logically structured. The information presented is relevant and substantiated with the use of examples.

Level 2 (3–5 marks)	The candidate will show good knowledge and understanding of some of the ecological issues surrounding the use of foam board. There will be some consideration of the environmental connection to these issues in relation to the lifecycle of a product or environmental impact. There will be limited reference to evaluating the issues. Evaluations will be one sided or limited to one factor. There is a line of reasoning presented with some structure. The information presented is in the most part relevant and supported by some evidence. Maximum of 4 marks if no evaluation evident.
Level 1 (1–2 marks)	The candidate will show limited knowledge and understanding of the ecological issues surrounding the use of foam board. There is no attempt at evaluation. The information has some relevance and is presented with limited structure or detail. The information is supported by limited evidence.
Level 0 (0 marks)	No response or no response worthy of credit.

Example 2: Timber-based materials

2 A storage unit made from chipboard is shown to the right.
  a) State two advantages of using chipboard for the storage unit compared to natural timbers. [2]
  b) The storage unit is assembled using knock down fittings.
     Name the two different types of knock down fittings shown below. [2]

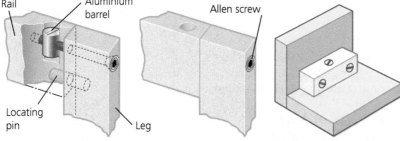

  c) State **two** advantages of using knock down fittings. [2]
  d) The chipboard is covered in a thin layer of plastic (melamine).
     Give **two** reasons why the chipboard is covered in the plastic (melamine). [2]

Candidate response

2 a) Chipboard is less expensive than natural woods.
     Chipboard is available in large sheets.
  b) Cam lock
     Corner block
  c) Easy to assemble because no specialist skills or tools are required.
     Can be taken apart so unit can be dismantled easily if needed.
  d) To stop it getting chipped and damaged.
     To make it waterproof and easy to wipe clean.

Assessment comment

a) One mark for each correct advantage up to a maximum of two.
b) One mark for identifying each fitting.
c) One mark for each correct advantage of knock down fittings up to a maximum of two.
d) One mark for each correct reason.

Example 3: Metal-based materials

3 a) A mild steel bar needs to be annealed to make it easier to bend into shape. Describe the annealing process. [3]

b) Complete the table below to show how each tool would be used when drilling holes in mild steel bar. [2]

Tool	What it is used for
Scriber	
Centre punch	
Hand vice	Holding the metal securely and safely while it is being drilled

c) Other than welding, name one other method of permanently joining pieces of mild steel together. [1]

Candidate response

3 a) The metal is heated gradually until it glows red. It is then allowed to cool slowly.

b)

Tool	What it is used for
Scriber	Used to mark out the hole position on the steel bar
Centre punch	Used to punch an indent in the centre of the hole so the drill does not 'wander'
Hand vice	Holding the metal securely and safely while it is being drilled

c) Brazing.

Assessment comment

a) One mark for each relevant point. The candidate gains marks as follows:
The metal is heated gradually [1] until it glows red [1]. It is then allowed to cool slowly. [1]

b) One mark for explanation of how the tools are used.

c) One mark for any appropriate method of joining steel.

Example 4: Polymers

4 a) Polypropylene is used to make a clear plastic bottle for mouthwash. State **three** qualities of polypropylene that make it suitable for this application. [3]

b) State the manufacturing process used to make a clear plastic bubble for the toothbrush packaging. [2]

c) Use sketches and notes to describe the manufacturing process used to make the clear plastic bubble. [7]

Candidate response

4 a) Transparent so you can see the product.
A thermoplastic that can easily be moulded by blow moulding.
Will not let contamination pass through.

b) Vacuum forming.

c) Vacuum forming is used to melt plastic into a different shape. The stages are:
Fasten the plastic on to the machine and heat it up.
Pull up the handle.
Switch on the vacuum so it sucks it into shape.
Remove the plastic sheet from the mould.
Cut off the extra plastic.

Assessment comment

a) All three answers describe properties of polypropylene that make it suitable for the clear plastic bottle. One mark for each answer.
b) Vacuum forming – one mark for each word.
c) One mark for each stage in the vacuum-forming process described in sketches and/or notes:
Mould placed on the machine.
Plastic sheet clamped in place.
Sheet heated until soft.
Mould raised.
Vacuum switched on to suck out air.
Forming removed from the mould.
Forming trimmed to size.

Example 5: Textile-based materials

5* A designer for a high street chain is making a new dress.
The dress will be made from 100 per cent cotton.
Discuss the ecological issues that the designer must consider. [8]

Candidate response

Cotton is the most widespread, profitable, non-food crop in the world. Its production provides income for more than 250 million people worldwide and employs almost seven per cent of all labour in developing countries. Approximately half of all textiles are made of cotton. The global reach of cotton is wide, but current cotton production methods are environmentally unsustainable. Brazil, Pakistan, Turkey and the USA are the world's leading producers of cotton crops. Cotton is intensively farmed across huge areas of land using fertilisers, pesticides and large amounts of water. The fibre is harvested from the seed pod (boll) of the cotton plant. A single mature cotton plant can produce around 100 cotton bolls annually. This roughly equates to one bale (225 kg). To put this into perspective, it takes around 1 kg of cotton to make a single pair of jeans. Over one billion pairs of jeans are produced worldwide, each year.

The cotton harvest begins in July and continues until November. Cotton picking can be completed by hand or machine, depending on the scale of production on the farm – machinery almost always contributes to environmental air pollution.

Most cotton is used to produce yarns for the manufacture of fabric. For this, it must be cleaned and bleached to achieve consistency in the finished product. The cotton fibre goes through several stages of cleaning before it can be spun into yarn. Chemicals soak through the soil and into water sources, killing wildlife and contaminating drinking water. These chemicals also strip the soil of nutrients, reducing crop quality and eventually making the land unusable.

Organic farming reduces this impact through the use of natural fertilisers and pesticides – this type of farming, still requires a substantial amount of water, however. It is estimated that approximately 2700 litres of water is required to produce enough cotton for a single T-shirt. There are few countries where fibre production is local to its manufacturers. Transportation of raw fibres causes air pollution and uses a significant amount of non-renewable resources.

Assessment comment

This is a level three response but it would not quite achieve full marks. There is a lot of factual and detailed information which demonstrates that the candidate has a very good understanding of the ecological issues surrounding the life cycle of cotton, but, the answer is constructed into a series of statements rather than a reasoned response to the question. The candidate does not refer to the production of the cotton dress or relate the answer to what issues the designer must consider, it appears to be very general information about cotton production. Candidates must always try to ensure that they address the issues in the stem of the question, they should try to debate the advantages and disadvantages of the issue, including examples and using technical terms where possible.

Mark scheme

Level 3 (6–8 marks)	The candidate will show good knowledge and understanding of the ecological issues that arise with the use of cotton. There should be good consideration of the environmental connection to these issues in relation to the lifecycle of a product or environmental impact. They will be able to undertake a thorough evaluation of the issues identifying positive and negative implications. There is a well-developed line of reasoning which is clear and logically structured. The information presented is relevant and substantiated with the use of examples.
Level 2 (3–5 marks)	The candidate will show good knowledge and understanding of some of the ecological issues surrounding the use of cotton. There will be some consideration of the environmental connection to these issues in relation to the lifecycle of a product or environmental impact. There will be limited reference to evaluating the issues. Evaluations will be one sided or limited to one factor. There is a line of reasoning presented with some structure. The information presented is in the most part relevant and supported by some evidence. Maximum of 4 marks if no evaluation evident.
Level 1 (1–2 marks)	The candidate will show limited knowledge and understanding of the ecological issues surrounding the use of cotton. There is no attempt at evaluation. The information has some relevance and is presented with limited structure or detail. The information is supported by limited evidence.
Level 0 (0 marks)	No response or no response worthy of credit.

Example 6: Design engineering

6 a) Light-emitting diodes (LEDs) are often used instead of bulbs in many modern electronic devices. Describe **two** advantages and **one** disadvantage of using LEDs instead of bulbs in electronic devices. [4]
   b) Mechanisms are used to transfer motion from one type to another.
   State the **two** types of motion used in the mechanisms below:
   i) Eccentric cam
   ii) Rack and pinion. [2]

Candidate response

6 a) Advantage 1 – LEDs last much longer than light bulbs.
   Advantage 2 – LEDs use much less energy than light bulbs.
   Disadvantage – LEDs only allow current to pass through in one direction so only work if connected correctly.
   b) i) Eccentric cam: Rotary motion into reciprocating motion.
      ii) Rack and pinion: Linear motion into rotary motion or vice versa.

Assessment comment

6 a) One mark for each correct advantage up to a maximum of two, and one mark for a correct disadvantage.
   b) One mark for each type of motion.

Glossary

5WHs: A way of identifying the needs and wants of the primary user and stakeholders by asking who, what, when, where and why.

Aesthetics: Factors concerned with the appreciation of beauty – this can include how something looks, sounds, feels, tastes and smells.

Alloys: Metals mixed with other elements to improve their characteristics, for example hardness.

Amplify: Make larger.

Annealing: A heating process applied to metal to increase its ductility and reduce hardness.

Anthropometrics: The study of the sizes of the human body.

Autonomous: Acting alone.

Batch production: Batch-produced products are manufactured by a large team of workers who each complete a specific stage of the production.

Bespoke production: Manufacture of 'one-off' products that are designed and made for a specific client by an individual or small team of highly skilled workers. Bespoke products are high quality, can be complex and are expensive to make.

Biobased: A product from a renewable source.

Biodegradeable: The ability of a substance to break down naturally in the environment through the actions of micro-organisms.

Brittle: A material that will shatter or break rather than bend and deform when forces are applied.

CAD: Computer-aided design.

CAM: Computer-aided manufacture.

Capabilities: The abilities of a product to complete a task.

Cellulose: Wood fibres, an organic compound structurally important in plant life.

CNC: Computer numeric control.

Collaboration: Working with others for mutual benefit.

Compatible: How a product fits or works with other products.

Composted: Naturally broken down.

Conductivity: How easily electricity, heat or sound is transmitted through a material.

Consumed: Eaten.

Convert: change from one thing, or state, to another.

Corrosion: The degradation of a material from elements such as oxygen, moisture and other chemicals.

Cultural factors: The beliefs, moral values, traditions, language, laws and behaviours that are common to a group of people (for example, a nation or community).

Decompose: To decay and become rotten.

Deforestation: The over-harvesting of trees creating areas of bare land.

Density: The mass of a material (its weight) divided by its volume (size).

Design solution: Where a product or system fulfils a need or want.

Dimensions: Measurements.

Disassembly: The taking apart of a product.

Distortion: A false view.

Ductility: How easily a material can be deformed or bent out of shape without snapping or breaking.

Durability: The ability of a material to withstand wear, pressure or damage.

Economic factors: How the making, using and disposing of products and services can have an impact on the industry and trade of a country.

Electrolysis: A method of extracting metal using an electric current.

Ellipse: Also called an oval.

Embedded: A program in an appliance that cannot be altered by the user, for example a washing machine.

Environment: The natural and man-made world around us.

Ergonomic: Concerned with the physical shape of an object.

Ergonomics: The study of how we use and interact with a product or system.

Ethical awareness: Concerned with doing the right thing.

Ethical: Correct, good or honourable. Aim for an ethical approach when you are designing products.

Extruded: A shape or material that has been produced by forcing it through a die.

Eye line: The height of the eyes.

Fast fashion: A recent trend involving the quick transfer of new collections from the catwalk into shops. Fast fashion is often on-trend, low quality and low in price. The characteristics of fast fashion mean that consumers buy large volumes of clothes more regularly, creating more profit.

Ferrous metals: Metals that contain iron.

Fibres: A thread or filament from which textiles are formed. They are tiny hair-like structures that are spun (twisted) together to make yarns. These yarns are then woven or knitted together to create fabric.

Finite: Limited in supply.

Flexible: Bends easily without snapping.

Fluted: A shape or object that has grooves or ridges.

Friction: resistance caused by two surfaces rubbing together.

Fulcrum: – the point around which a bar or beam rotates.

Gear ratio: The diameter or number of teeth on the driven divided by the driver.

Gear train: Two or more gears meshed together.

Globalisation: The process by which the world is becoming increasingly interconnected as a result of massively increased trade and cultural exchange.

Golden ratio: A common mathematical ratio found in nature that can be used to create pleasing, natural-looking designs.

Grain line: The grain line always follows the direction of the warp yarn, and shows where the pieces should be placed on the fabric prior to cutting.

gsm: Grams per square metre. Used to classify the weights of paper and card.

Hardwood: Wood that comes from deciduous trees.

Holographic: A special type of photograph or image in which the objects look three dimensional rather than two dimensional.

Horizontal: On its side or level with the horizon.

Idler gear: An extra gear used to keep the driver and driven gear turning in the same direction.

Inclusive design: The design of mainstream products and/or services that are accessible to, and usable by, as many people as reasonably possible without the need for special adaptation or specialised design.

Increase: Make larger.

Input device: A device that senses something, for example temperature.

Insulation: A material that prevents heat, electricity or sound from escaping.

Just-in-time manufacturing: A manufacturing method that can quickly respond to changes by ordering materials and components to arrive at the factory 'just in time' for production.

Knock down fittings: Fittings used in flat-pack furniture to make assembly easy.

Landfill: The disposal of waste matter by burying it.

Lay plans: Used by pattern cutters like a map to guide them when placing pattern pieces in the correct location and direction before cutting. Lay plans are carefully designed to minimise wastage.

Lean manufacturing: A manufacturing method designed to minimise waste at each stage of production.

Lifecycle assessment: The analysis of the impact of a product on the environment throughout the manufacture, use and disposal of that product.

Lightweight: Weighs very little.

Lubrication: A substance applied to reduce friction.

Malleability: A material's ability to deform without breaking or snapping when hammered or rolled into a thin sheet.

Managed forest: A forest in which more trees are planted to replace every tree that is felled.

Manufactured board: Man-made boards made from recycled natural woods.

Marketing: The business of promoting and selling a product; it can include advertising and promotion, and market research.

Mass production: Mass-produced products are manufactured mostly on automated machinery, operated by teams of workers. Products are manufactured very quickly, are consistent in quality and the range of available styles is minimal. Mass production costs are much lower than other production methods.

Mechanical advantage: Using a mechanism to increase output force.

Microcontroller: A programmable electronic component.

Microencapsulation: The process of adding tiny bubbles or capsules that contain chemicals to the weave or fibre of fabric. These burst with friction, releasing the chemical inside.

Micron: One thousandth of a millimetre. Used to classify the thickness of paper and card.

Microprocessor: Electronic system used to control output devices.

Moral factors: A moral issue is related to human behaviour; it is the distinction between good and bad, or right and wrong, behaviour according to our conscience.

Motion: movement.

Natural cellulose fibres: Fibres that come from plant-based sources, for example, cotton and linen.

Natural polymers: Chains of protein or cellulose molecules (monomers), such as keratin, glucose or fibroin. These chains are the basis of all natural fibres.

Natural protein fibres: Fibres that come from animal-based sources. These include hair, fur or silk fibres.

Natural woods: Woods that come from trees.

Non-biodegradable: Does not degrade or rot down into a harmless state.

Non-ferrous metals: Metals that do not contain iron.

Opaque: Does not allow light to pass through.

Output device: A device that produces audible or visual indicators, or movement.

Photochromic: Products that change colour with changes in natural light.

Plasticity: The ability of a material to permanently change in shape when force is applied to it.

Porous: A material that has many tiny holes that allow moisture to seep through.

Primary user: The main user of the product or system.

Program: A set of instructions loaded on to a microcontroller.

Prototype: A test model on which the final product is based.

Quality assurance: Consistent quality testing at each stage of production means that manufacturers can maintain the required level of quality in their products.

Quality control: Sample products are taken off the production line at critical control points to check for faults or to test performance.

Reclaimed timber: Old timber that is reused in a new way.

Recycled paper: Paper made from used paper products.

Reduce: Make smaller.

Refined: Minor changes to improve the product.

Replenished: Quickly restored to previous levels.

Rigid: Difficult to bend.

Seasoning: Carefully drying woods ready for use.

Signal: An electronic voltage that is used to represent information.

Social awareness: How the designing, making and use of products impact upon people.

Social factors: Lifestyle factors that affect people within our society.

Softwood: Wood that comes from coniferous (evergreen) trees.

Stakeholder: A user, person, group or organisation that has an interest in the product or system.

Synthetic fibres: Man-made fibres that come from a range of sources including coal, oil, minerals and other petrochemicals.

Systems thinking: The understanding of a product or component as part of a larger system of other products and systems. In the iterative design process, consideration of the role of all components and subsystems of the product or system, including the user experience and the marketing of the object being designed, ensures all aspects of the product are given the required attention to detail.

Task analysis: An exploration of the design context.

Tempering: A heating process applied to metal to make it less brittle.

Tenacity: The strength of textiles and fabrics.

Thermochromic: Products that change colour with changes in temperature.

Throwaway culture: The rise of fast fashion has made clothing more affordable for consumers. The incredibly low prices in some high street stores have resulted in a throwaway culture, meaning that consumers don't feel the need to keep clothing that is no longer in fashion, and happily dispose of it.

Throwaway society: A society influenced by consumerism and excessive consumption of products.

Torque: Rotational force.

Toxic: Harmful.

Transparent: See-through.

Triangulation: Adding strength to frames by adding triangle shapes.

Usability: How easy a product is to use, how clear and obvious the functions are.

User-centred design (UCD): Sometimes called 'human-centred design', the user-centred design strategy aims to make products and systems useable. It focuses on the user interface and how the user interacts with and relates to the product.

Vertical: Upright.

Virgin fibre paper: Paper made from 'new', unused wood fibres.

Virtual world: Not real, imaginary.